# Propagation of the Houbara Bustard

# PROPAGATION OF THE HOUBARA BUSTARD

*EDITED BY*
M. SAINT JALME and Y. VAN HEEZIK

*ART DIRECTOR*
P. PAILLAT

*PHOTOGRAPHS BY*
X. EICHAKER
AND THE STAFF OF THE NATIONAL WILDLIFE
RESEARCH CENTER

KEGAN PAUL INTERNATIONAL
London and New York

JOINTLY WITH THE

NATIONAL WILDLIFE RESEARCH CENTER
NATIONAL COMMISSION FOR WILDLIFE
CONSERVATION AND DEVELOPMENT
Riyadh, Saudi Arabia

First published in 1996 by
Kegan Paul International
UK: P.O. Box 256, London WC1B 3SW, England
Tel: (0171) 580 5511 Fax: (0171) 436 0899
E-mail: books@keganpau.demon.co.uk
Internet: http://www.demon.co.uk/keganpaul/
USA: 562 West 113th Street, New York, NY 10025, USA
Tel: (212) 666 1000 Fax: (212) 316 3100

Distributed by

John Wiley & Sons Ltd
Southern Cross Trading Estate
1 Oldlands Way, Bognor Regis
West Sussex, PO22 9SA, England
Tel: (01243) 779 777 Fax: (01243) 820 250

Columbia University Press
562 West 113th Street
New York, NY 10025, USA
Tel: (212) 666 1000 Fax: (212) 316 3100

© National Commission for Wildlife Conservation and Development 1996

Phototypeset in Garamond by Intype London Ltd

Printed in Great Britain by TJ Press, Padstow, Cornwall

All rights reserved. No part of this book may be reprinted
or reproduced or utilized in any form or by any electronic,
mechanical or other means, now known or hereafter invented,
including photocopying and recording, or in any information
storage or retrieval system, without permission in writing
from the publishers.

ISBN 0-7103-0518-4

*British Library Cataloguing in Publication Data*

Propagation of the Houbara Bustard
I. Jalme, M. Saint II. Heezik, Y. Van
639.97831
ISBN 0-7103-0518-4

*US Library of Congress Cataloging in Publication Data*

Propagation of the houbara bustard / edited by M. Saint Jalme and Y. van Heezik.
112 pp. 24 cm.
Includes bibliographical references and index.
ISBN 0-7103-0518-4
1. Houbara—Breeding—Saudi Arabia. 2. Houbara—Saudi Arabia.
I. Saint Jalme, M. II. Van Heezik, Y.
SF510.H68P76  1995
639.9'7831—dc20                                                    95-14864
                                                                    CIP

*In the Name of God,
Most Gracious,
Most Merciful.*

# Contents

*List of Illustrations*  ix
*List of Tables*  xi
*Foreword*  xiii
*Preface*  xv
*Acknowledgements*  xvii

| | | |
|---|---|---:|
| 1 | Introduction<br>P. Seddon | 1 |
| 2 | Biology and Status of the Houbara Bustard<br>H. Schulz and P. Seddon | 3 |
| 3 | Management and Facilities<br>P. Paillat and P. Gaucher | 15 |
| 4 | Reproductive Parameters in Captivity<br>M. Saint Jalme, P. Gaucher and P. Paillat | 32 |
| 5 | Artificial Insemination and Natural Mating<br>P. Gaucher, M. Saint Jalme and P. Paillat | 45 |
| 6 | Incubation<br>P. Gaucher, P. Paillat and M. Saint Jalme | 57 |
| 7 | Rearing<br>P. Paillat and Y. van Heezik | 65 |
| 8 | Pathology and Veterinary Care<br>S. Ostrowski, A. Greth and I. Mikaelian | 70 |
| 9 | Production and Prospects<br>M. Saint Jalme, P. Paillat and P. Gaucher | 97 |
| 10 | Conclusions<br>P. Seddon | 100 |

*References*  103

# List of Illustrations

**Plates** *(between pages 14 and 15)*

2.1 The Houbara Bustard
2.2 *Chlamydotis undulata undulata*
2.3 *Chlamydotis undulata macqueenii*
2.4 Typical Habitat of Houbara Bustard
2.5 Houbara Bustard Nest
2.6 Houbara Chick
3.1 Aerial View of the Houbara Breeding Unit
3.2 A Block of Cages in the Breeding Unit
3.3 Inside a Cage in the Breeding Unit
4.1 Displaying Male Houbara
5.1 Female Houbara Passing Through a Size-selection Door
5.2 Pre-copulatory Display Directed Towards a Dummy Female
5.3 Semen Collection and Artificial Insemination Equipment: Insemination Syringe, Aspiration Apparatus, Diluent, Dish, and Forceps Used to Open the Cloaca
5.4 Collecting Equipment Used to Aspirate the Semen from the Cloaca into a Vial Held in the Operator's Hand
5.5 Insemination Procedure
6.1 The Incubation Room
7.1 Female Feeding a Chick
7.2 Plexiglas Brooding Boxes Containing Young Chicks
7.3 Hand-feeding a Cricket to a Chick
7.4 Hand-feeding Chicks
8.1 Presence of a Foreign Body (Screw) in the Ventriculus of a 1-week-old Chick (X-ray)
8.2 Vaccination of an Adult Houbara
9.1 Houbara Chicks
10.1 One-week-old Houbara Chick
10.2 Released Houbara in Flight in Mahazat as-Sayd
10.3 Houbara on Nest in Harrat al-Harrah

*List of Illustrations*

10.4 Houbara Chick in the Wild

## Figures

| | | |
|---|---|---|
| 2.1 | Houbara Distribution | 5 |
| 3.1 | Layout of Houbara Breeding Pens | 19 |
| 3.2 | Layout of Part of a Block Within the Breeding Unit | 20 |
| 3.3 | Layout of the Incubation Complex | 22 |
| 3.4 | Layout of Air Intake to the Air-conditioning Room | 24 |
| 3.5 | Layout of the Incubation Room | 25 |
| 3.6 | Layout of the Rearing Unit | 27 |
| 3.7 | Handling Technique | 28 |
| 4.1 | Seasonal Variations in the Number of Laying Female Houbara, Precipitation and Maximum Ambient Temperatures, 1990–1993. | 35 |
| 4.2 | Annual Variation in Male Houbara Body Mass, Number of Displaying Males and Rate of Moult, i.e. Number of Primaries Simultaneously in Growth | 37 |
| 4.3 | Annual Variation in Female Houbara Body Mass, Number of Laying Females and Rate of Moult, i.e. Number of Primaries Simultaneously in Growth | 38 |
| 5.1 | Male Mounting a Dummy Female During the Semen Collection Procedure | 49 |
| 5.2 | Effect of Sperm Storage *in Vivo* on Fertility and Embryo Mortality in Houbara | 53 |
| 6.1 | Apparatus for Recording Incubation Parameters | 59 |
| 6.2 | Example of Mass-loss Recording Sheet | 60 |
| 7.1 | Growth Curves of Male and Female *macqueenii* Chicks | 68 |
| 8.1 | Causes of Death Among (a) Houbara Between 1987 and 1991 (b) Houbara in 1993 | 71 |
| 8.2 | Mean Doxycycline Blood Concentrations Versus Time After the First Injection, During the Initial Seven-day Period: Intramuscular and Subcutaneous | 78 |
| 8.3 | Mean Doxycycline Blood Concentrations Versus Time After the Seventh Injection, During the Next Seven-day Period: Intramuscular and Subcutaneous | 79 |

# List of Tables

| | | |
|---|---|---|
| 3.1 | Origin of Birds in the NWRC Breeding Flock | 16 |
| 3.2 | Age of the Breeding Flock Between 1988 and 1993 | 17 |
| 3.3 | Management of the Breeding Flock Between 1988 and 1993 (Birds Aged Two Years and Over) | 18 |
| 4.1 | Date of First Egg Laid | 34 |
| 4.2 | Relationship Between Age and Laying (Percentage of Laying Females in Five Age Classes) | 39 |
| 4.3 | Percentage of Laying Females and Production of Eggs in Each Age Class in 1993 | 40 |
| 5.1 | Summary of Natural Breeding Results, 1990–1993 | 51 |
| 5.2 | Results of Artificial Insemination, 1989–1993 | 52 |
| 5.3 | Relationship Between Number of Spermatozoa Inseminated During Three Time Intervals Before Laying, and Fertility Level of Eggs as a Percentage of Total Eggs | 54 |
| 6.1 | Embryo Mortality | 62 |
| 6.2 | Comparison Between Levels of Embryo Mortality Associated with Two Incubation Techniques | 63 |
| 8.1 | Macroscopic Lesions in Houbara Showing Inclusion Bodies Resembling *Chlamydia* | 75 |
| 8.2 | Prevalence of Antibodies for Thirteen Microbial Pathogens in Houbara | 86 |
| 8.3 | Drugs and Dosages Used for Houbara | 94 |

# Foreword

HIS ROYAL HIGHNESS PRINCE SAUD AL FAISAL
*Minister of Foreign Affairs*
*Managing Director*
*National Commission for Wildlife Conservation and Development*

Since the formation of the Kingdom of Saudi Arabia, the Government has shown a strong desire to preserve and protect its unique traditions for the benefit of its citizens. The fast pace of the Kingdom's development has produced positive effects in many sectors, such as agriculture, education and social services. However, the price of progress can often be costly in terms of the degradation of natural habitats leading to the decline of wildlife resources. Now that Saudi Arabia has stepped resolutely into the modern world, the wounds inflicted to our natural environment should be healed by restoring the traditional relations of man with nature.

This book will highlight the achievements accomplished in the short span of seven years by the staff of the National Wildlife Research Center in laying the foundation for environmental restoration.

Thanks to Professor Abdulaziz H. Abuzinada, Secretary General of the National Commission for Wildlife Conservation and Development, and to the dedicated research team headed by Jacques Renaud, the houbara propagation programme started in 1986 at the National Wildlife Research Center in Taif. At this time little was known about the houbara's environment and biology. Today this is no longer the case as we have accumulated much data through research and observations, enabling the conservation and development of this endangered species. We now know the houbara will not be destined for the history books but restored once again to take its rightful place within Saudi Arabia.

The captive breeding and reintroduction of the houbara bustard is the first step in a comprehensive programme of environmental protection and restoration of Saudi Arabia's biodiversity. However, scientific research alone does not guarantee the success of such an enterprise, and the active support of the public should

## Foreword

be raised by extending public awareness and environmental education in general. Our hopes are that the work so far accomplished by the National Wildlife Research Center will arouse the greatest interest among the young Saudi generation.

# Preface

PROFESSOR ABDULAZIZ H. ABUZINADA
*Secretary General*
*National Commission for Wildlife Conservation and Development*

It gives me great pleasure to contribute to the preface of this book presenting the output of our very first conservation programme.

For centuries the houbara has beguiled the hearts and minds of people of the Arabian Peninsula. Traditional hunting with falcons for houbara was an integral part of life in the desert. Among early travellers and naturalists the houbara was admired and known for its spectacular display. Over the last few decades this species, so well adapted to desert environments, has suffered greatly with many others at the hands of man. Today the future of the houbara bustard is at stake, due to continuous degradation of its habitat by domestic livestock, and to increasing hunting pressure, even in its last retreats.

These continuous threats and the poor prospects for the species in the wild led to the establishment of the National Wildlife Research Center. Initiated by His Royal Highness Prince Saud Al Faisal and executed by Mr Jacques Renaud, the initial goal was to secure the survival of the species through captive breeding.

As the book documents, this goal has now been reached by overcoming the many hurdles that accompany such operations. I hope that the success we have achieved in understanding the many enigmas of houbara breeding biology and physiology can now be transferred into a successful restocking and reintroduction programme, ultimately to restore this magnificent bird in the land of Arabia.

Today, the houbara bustard programme represents one of the pioneering projects that created the necessary momentum and motivation for the achievements that have been made by the National Commission for Wildlife Conservation and Development in protecting and restoring the country's biodiversity. It is very encouraging to see the efforts that are currently underway

## Preface

by the National Wildlife Research Center to assist us in this challenging time.

As the Secretary General of the National Commission for Wildlife Conservation and Development, I witnessed the project since its inception, and it is very fulfilling to see the comprehensive scientific report presented here today. The information provided here will be instrumental in establishing our management policies for conservation of houbara bustard in Saudi Arabia.

# Acknowledgements

The concept of this project has been supported by His Royal Highness Prince Sultan bin Abdulaziz Al Saud, Chairman of the Board of Directors of the National Commission for Wildlife Conservation and Development.

We thank His Royal Highness Prince Saud Al Faisal, Managing Director of the National Commission for Wildlife Conservation and Development, who initiated this project and whose continued interest has been a great support. We are also grateful to Professor Abdulaziz H. Abuzinada, Secretary General of the National Commission for Wildlife Conservation and Development for his guidance and support and to Dr Saud Al Shawaf for his invaluable assistance.

Drs M. de Reviers and B. Sauveur shared their knowledge of captive breeding with us, Dr T. Burke performed the fingerprinting studies, and Drs B. Andral, H. Gerlach and H. Gerbemann contributed to our understanding of houbara pathology.

Many people have participated enthusiastically in this project during the last eight years: we are grateful to A. Khoja, Administrative Director of the National Wildlife Research Center, for his patience, kindness and efficiency, and to the following staff: Dr A. Ancel, Dr S. Anagariyah, K. Anegay, J.F. and A. Asmodé, Dr S. Biquand, Dr V. Biquand-Guyot, A. Boug, Dr O. and F. Combreau, Dr S. Darroze, V. Dorval, Drs L. and J. Durand, X. Eichaker, Dr J. and I. Glamand, Dr H. Gillet, Dr L. Granjon, D. Le Mesurier, F. and C. Launay, P. McCormick, Drs S. and A. Newton, Dr T. Petit, Drs J.F. and A. Poilane, Drs R. and J. Seitre, M. Shobrak, P. and P. Symens, A. Vareille, Dr M. Vassart, Dr A. Verdier, C. Weigelt and C. Wilme. Thanks also to Dr S. Biquand for his input into the first draft of this book, and to reviewers Mr D. Hancock and Dr T. Cade, for their constructive criticisms and helpful suggestions.

# 1
# Introduction

P. SEDDON

The National Wildlife Research Center (NWRC), near Taif, Saudi Arabia, was established in 1986 with the single aim of breeding houbara bustards in captivity. Now, nine years later, the NWRC has expanded its work to include the captive-breeding and reintroduction of other endangered Arabian animals, but the houbara programme remains the core project.

The houbara at first glance appears an unlikely example of a flagship species around which the conservation efforts of Saudi Arabia have grown. It is cryptic to the point of invisibility, with patterns of brown and black blending perfectly with the desert. But, in flight, or in nuptial display, the houbara blooms black and white, visible over long distances across treeless landscapes; a beacon for the falcons which are trained to hunt it. It is as the foremost quarry for Arab falconers that the houbara attains its importance in the Middle East.

Oil exploration, modernization, and the prevalence of four-wheel-drive vehicles since the 1940s have been associated with growing numbers of falconers and hunting parties in the Arabian Peninsula, and have greatly increased the impact falconry has had on houbara populations. Hunting, together with loss of habitat through agricultural development, has reduced the numbers of houbara throughout its range. It was in recognition of the decline of the houbara in Saudi Arabia that His Royal Highness Prince Saud Al Faisal established the NWRC. The goal of the houbara breeding programme at Taif was the production of a self-sustaining captive population of houbara, and the provision of an annual surplus of houbara chicks for rearing and release back into suitably protected sites in the Kingdom.

The houbara breeding programme began in the shadow of possible failure. The lack of basic information about the bird's biology made captive-breeding a daunting task. However, the breeding programme has proven to be the greatest source of information about the reproductive biology of the houbara, yielding data not obtainable at present from wild houbara. The early stages of the Taif project were very much a process of trial and error, and occasionally, trial and success. The experience that was gradually gained was fed back into the management of the captive flock with the result that in 1989 the first captively bred houbara chick was hatched. Success built upon success, and by 1992 the captive flock was self-sustaining, and the first 'soft release' trials had begun in the nearby Mahazat as-Sayd protected area, a fenced reserve of about 2,200 km$^2$.

This book is presented as a 'how to' manual for breeding houbara in captivity, but it is also a little more. It presents and summarizes the background data obtained on houbara diet, behaviour, physiology and veterinary care that have enabled the breeding programme to succeed. In this way it is intended to be of interest not only to those attempting to keep or breed houbara, but also to those who are studying or managing wild populations. In addition, there is perhaps something here of interest to keepers or breeders of other bustard species.

Captive-breeding is in itself not the answer to halting the decline of a species. Concurrent efforts must be made to remove the causes of the decline and to protect those populations that remain in the wild. These concerns are being addressed by Saudi Arabia's conservation authority, the National Commission for Wildlife Conservation and Development (NCWCD). Captive-breeding of the houbara, however, has provided the results in the hand necessary to maintain the support for wider conservation efforts for the houbara, and for other wildlife in Saudi Arabia.

# 2
# Biology and Status of the Houbara Bustard

H. SCHULZ and P. SEDDON

## 2.1 Taxonomy and Morphology

The houbara (*Chlamydotis undulata*) is a medium-sized bustard of slender appearance, measuring 55–65 cm. and having a wingspan of 135–170 cm. The general colour of the houbara is sandy buff, spotted with blackish vermiculations all over the upper body. The neck, sides of the face and ear coverts are pale sandy buff, with a black stripe pattern created on both sides of the neck by elongated display feathers (Plate 2.1). The underparts, in contrast, are mainly white. In flight, its wings exhibit a black and white pattern. The body mass of adult birds in captivity varies from 1,200 to 2,400 g. in males and from 1,100 to 1,700 g. in females.

The houbara is the only species of the genus *Chlamydotis*, being separated into three sub-species: the nominate race *Chlamydotis undulata undulata* (Plate 2.2), and the sub-species *C. undulata macqueenii* (Plate 2.3) and *C. undulata fuertaventurae*. The sub-species differ in colouration and size, *macqueenii* being the largest and palest and *fuertaventurae* being the smallest and darkest. Major differences are in the extent and size of the dark vermiculations of the body plumage, in the colouration of the tuft of feathers on the crown, which is pure white in *undulata*, and which has a black tip in *macqueenii*, and in the filamentous display plumes along the side of the neck, which are all black in *undulata* and are black and white in *macqueenii*. In all the three sub-species the female is smaller than the male, and has shorter crest and neck feathers.

Differences in courtship displays between *undulata* and *macqueenii* also exist. The courtship display of the male houbara is striking: by erecting normally inconspicuous feathers and adopting patterns of movement that can be likened to a stylized dance, the male birds perform an extremely conspicuous visual display. Differences between the sub-species are as follows:

1. The position and colour of certain feathers used during the display: the frills on both sides of the neck are completely black in *undulata*, but are black and white in *macqueenii*. The crest is white and remains completely erected in *undulata* whereas it is black and white and falls over the bill in *macqueenii*.
2. During the running display *undulata* males tend to run faster than *macqueenii* males. Neck-swaying is of lower amplitude in *undulata*.
3. The vocalization at the end of the running display is a discrete call repeated two-four times in *undulata*, whereas it is a continuous call in *macqueenii*.

These differences which developed in geographic isolation may now allow for reproductive isolation in areas of sympatry, such as Egypt (Goodman & Meininger 1989), as results from artificial insemination indicate that no cytogenetic barriers exist to prevent the hybridization of *undulata* and *macqueenii*. This evidence, together with results from genetic studies of DNA and protein structure, suggests the two sub-species should be considered as two separate species (Gaucher *et al.* in press).

## 2.2 World Distribution

The breeding distribution of the houbara ranges from the Canary Islands through Morocco, Mauritania, Algeria and Tunisia to Libya, from Egypt to Palestine and the Arabian peninsula and from the Caspian Sea eastwards over the Aral Sea to 120° east into Mongolia (Fig. 2.1). From the Caspian Sea it extends southwards into Afghanistan, Pakistan, Iran and Iraq. The sub-species *fuertaventurae* only lives on the Canary Islands of Fuerteventura and Lanzarote, the nominate sub-species *undulata* inhabits the North African range, and *macqueenii* appears from Sinai eastwards, perhaps also at the eastern edge of the Nile valley, where it may be sympatric with *undulata*.

The houbara breeding in central Asia are migratory, whereas breeding populations of the Middle East, Arabia and Africa, and perhaps partially also those from Pakistan, are non-migratory,

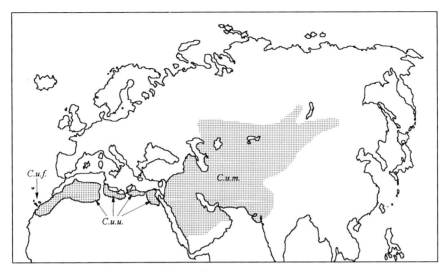

C.u.u. = *Chlamydotis undulata undulata*
C.u.m. = *Chlamydotis undulata macqueenii*
C.u.f. = *Chlamydotis undulata fuerteventurae*

**Figure 2.1 Houbara distribution**

but carry out short-range movements in response to varying food supply and rainfall. Autumn migration starts in late August, lasting till early November, and spring migration occurs during late March and April. Consequently, wintering flocks occur in Pakistan as early as September, but many more birds have arrived by November. In Saudi Arabia wintering flocks of houbara can be found between November and March.

After breeding, houbara of central Asia migrate in a south-western direction to North-west India (Rajastan, Gujarat), Pakistan, Southern Iran, Asia Minor, Arabia and Sinai, and occasionally possibly also to north-eastern Egypt and north-eastern Sudan. Details of the migration routes are unknown at present.

## 2.3 Breeding Distribution in Saudi Arabia

By all accounts it is clear that the numbers and distribution of houbara breeding in Saudi Arabia have undergone a dramatic decrease. Past records (reviewed in Porter & Goriup 1985) suggest that houbara once bred within an area extending from the Jordanian border in the north, down in a band encompassing the north-west and eastern regions as far south as the Rub-al-Khali. Recent reports suggest that the houbara is now an uncommon breeding visitor, restricted to north-western Saudi Arabia, specifically Harrat al-Harrah, Al Hammad and Al Nafud. Regular breeding by houbara in Saudi Arabia has been confirmed only for the Al Harrah region in the north-west (Green 1984; Symens 1988), where males may display as early as February, and eggs are laid between February and April.

## 2.4 Habitat

The houbara is adapted to desert environments, preferentially inhabiting undulating, flat arid plains, steppe habitats and semi-deserts, often with little cover except for open or scattered desert shrubs (Coles & Collar 1980; Mendelssohn 1980; Collar & Goriup 1983; Goriup 1983; Mian 1984; Alekseev 1985; Mian & Dasti 1985; Surahio 1985; Plate 2.4). Rugged terrain is avoided, as are pure sand massifs and barren salt pans (Ponomareva 1985). Vegetative cover consists of moderate or sparse perennials, primarily grasses, herbs and shrubs, but sometimes including larger bushes and trees. Common vegetative features from Pakistan and Israel include: desert vegetation, no regular tree layer, low absolute cover, shrubs widely spaced, some degree of loose soil, and all plants reproducing naturally despite persistent drought. Typical plant associations include *Artemisia*, *Haloxylon* and *Salsola* stands, cactoid *Euphorbias* and *Eremopterix* grasses. *Asphodelus microcarpus*, *Noaea mucronata*, *Zilla spinosa* and *Anabasis salsa* may also be present (Collar & Goriup 1983; Goriup 1983; Alekseev 1985; Malik 1985; Mian & Dasti 1985; Ponomareva 1985; Mian 1989). Wooded areas are completely avoided. On Fuerteventura houbara preferred sandy plains over stony areas, selecting areas richest in forbs (Collar & Goriup 1983). The generally darker plumage colour and less sandy appearance of *fuertaventurae* possibly has evolved in response to the blackish volcanic soil found on the Canary Islands.

Within their zone of distribution, annual rainfall rarely exceeds 200 mm. Where undisturbed, houbara may use cultivated areas (Mendelssohn *et al.* 1979; Lavee 1985; Razdan & Mansoori 1989) such as alfalfa fields (*Medicago sativa*), as was observed regularly on the Canary Islands and in birds kept in the Mahazat as-Sayd reserve in Saudi Arabia. In India they visit wheat fields and mustard cultivations.

During winter houbara are usually found in semi-arid to arid areas. In Harrat al-Harrah, they move between scattered food patches, foraging mainly on green vegetation in wadis and small silty depressions, but roosting in the elevated boulder fields at night (Seddon & van Heezik 1994).

## 2.5 Breeding Habitat

Houbara breed in arid or semi-arid zones receiving between 50 and 200 mm. annual rainfall, falling mainly in the winter (Mendelssohn et al. 1979; Haddane 1985; Mian & Dasti 1985). The substrate may be sand, gravel, or alluvial clays, and is usually loose, but not shifting, forming an even surface (Alekseev 1985; Mian & Dasti 1985; Mian 1989). In the breeding areas of Kyzylkum in Russia, houbara are found in alkaline alluvial clay landscapes vegetated by halophytes such as *Anabasis, Arthrophytum* and *Salsola*, as well as spring ephemerals that provide food for spring migrants (Alekseev 1985). Areas of pure sand are avoided in Kyzylkum, and nesting is concentrated on the *Anabasis*-dominated clay plains.

The breeding habitat of the houbara in Saudi Arabia follows the same general pattern of undulating lowland basalt plains, with loose sand or gravel substrate and scattered silt pans. The vegetation is scattered shrubs with associations of *Artemisia sieberi, Haloxylon salicornicum, Achillea fragrantissima, Zilla spinosa,* and *Astragalus spinosus* (Seddon & van Heezik 1995).

## 2.6 Food

Published accounts of houbara diet in Pakistan, Tunisia, Morocco, Oman and Russia agree that this species is omnivorous and opportunistic, the diet reflecting local and seasonal abundance of various plants and small animals (Ali & Ripley 1980; Coles & Collar 1980; Gallagher & Woodcock 1980; Goriup 1983; Mian & Surahio 1983; Mian 1984; Alekseev 1985; Surahio 1985). There appears to be some seasonal variation, with plants being a more important source of food during winter and early spring. Vegetable matter eaten includes fruits (e.g. *Ziziphus, Salsola, Lycium,*

*Launaea*), seeds, shoots, leaves, flowers and *Allium* bulbs, with young shoots, drupes, seeds, and berries preferred over leaves. Where available, cultivated plants are taken, such as beans, peas, alfalfa and mustard (Lavee 1985; Collar & Goriup 1983). Spring and summer foods are apparently more likely to include a variety of animals, including invertebrates such as grasshoppers, weevils, termites, locusts, beetles (Tenebrionidae, Scarabidae, Cantharidae), caterpillars, scorpions, spiders and ants; but also snails and small vertebrates such as snakes, lizards and geckos (Coles & Collar 1980; Collins 1980; Gallagher & Woodcock 1980; Goriup 1983; Mian 1984; Alekseev 1985; Gaucher 1991). Chicks are fed mainly on insects and small reptiles. Contradictory reports on drinking exist: houbara have been reported to drink regularly (Meinerzthagen 1954), whereas according to Dement'ev & Gladkov (1968) houbara never drink.

## 2.7 The Pair-bond

The houbara has been variously described as monogamous (Dement'ev & Gladkov 1968), polygamous and promiscuous (Collins 1984). Mendelssohn *et al.* (1979) report that pair bonds form in the spring, and that males are seen in the vicinity of females and chicks. In contrast, Dement'ev & Gladkov (1968) state that houbara pair during breeding, but that males are not seen near the nest. Some consider the species to be promiscuous, not forming any kind of pair-bond except during the short period of copulation. In Sinai the system appears to be variable: polygamous in high densities (e.g. in areas with many irrigated fields where food is abundant), but monogamous when population density is low.

In a promiscuous lek system, males occupy display territories which may be defended against other males. Houbara males have been observed frequenting display sites, which are usually found on open raised terrain with gentle slopes or low ridges (Goriup & Norton 1990). Males may display solitarily, although up to three males have been observed in display at distances of less than 200 m. to more than 1,000 m. from each other. However, it was not clear whether these birds were able to see each other (Goriup *et al.* 1992). In Algeria, Gaucher (1987) located two display grounds 2,300 m. apart, each situated 800 m. and 1,500 m. respectively from nest sites. Observations from the Canary Islands suggest that each male excludes other males from a territory of approximately 60 ha., held for feeding and roosting, within which is a specific display site (Collins 1984). Records from Kazakhstan state that fighting between males occurs (Dement'ev & Gladkov 1968). Displaying houbara in the Negev were not seen in close proximity to conspecifics (Mendelssohn et al. 1979). It is generally believed that the primary purpose of the male houbara display is attraction of a mate (Coles & Collar 1980). By displaying communally, males attract females which then have to choose a partner among neighbouring males by 'comparing' certain indicators of male quality, e.g. the duration of display runs. Females may then visit the male of their choice, copulate with him and immediately after move away to lay and incubate their eggs, some distance away from the males' display territories. Houbara males have never been seen incubating eggs or feeding chicks, suggesting that they are indeed truly promiscuous.

## 2.8 Display

At the beginning of the display sequence the male stands motionless; after some seconds, the elongated crown and neck-plumes are raised slowly up and forward, the wing-tips are lifted up slightly, and the tail is pressed down. The head is drawn back until it rests on the bird's shoulders, and the breast and neck-feathers continue to rise, finally entirely covering the head. The bird, which was previously well camouflaged, suddenly becomes visible as a white spot over distances of at least two kilometres in open land; it now looks like a bird with a large white ball attached to its breast.

After remaining in this posture for several seconds, the male begins to run in a straight line, in circles, or in a zigzag pattern, depending on the topography of the display ground and on the subspecies. He runs with short steps, lifting the legs in an exaggerated manner. The male abruptly stops at the end of each run, jerking its head and neck forward once, uttering a low-pitched noise and then pulling its head and neck back again. The bird may then move back into its motionless starting posture, and relax and return to normal, or it might stand with erect feathers until the next display run.

Most displays can be observed in the early morning and late afternoon. During the peak of the display season the display runs are repeated again and again, with breaks of only a few seconds to several minutes between them. Up to 31 displays were recorded in one display sequence, which lasted 70 minutes. During such sequences males do not or only very rarely feed. Display runs can be restricted to an area of only a few metres' diameter, but sometimes several hundred metres can be covered during consecutive display runs.

When a female appears at the display ground, the male begins a pre-copulatory display; it lowers the breast-feathers, approaches to directly in front of the female, and with the neck feathers still erected, swings its head and neck with forward-jerking movements, rapidly from right to left at about one-second intervals. When the female squats down, the male moves behind her, pecks her head and neck for several seconds and finally mounts her for copulation. The entire behavioural repertoire of the houbara has been described by Launay & Paillat (1990).

## 2.9 Incubation and Chick Rearing

Although the nesting season of the houbara is quite variable across the species' range, most clutches are produced during the spring months. Houbara nests are shallow scrapes occasionally lined with vegetation, that are situated on gentle slopes and elevated ground, rather than in depressions (Plate 2.5). The slopes tend to be south-facing, presumably to maximize isolation. The placement of the nest scrape with regard to other features varies with habitat. In Harrat al-Harrah, Saudi Arabia, nests have been found close to sandy wadis, in areas of small basaltic boulders and in good vegetative cover (Symens 1988). In Pakistan and Algeria nests were sited in areas clear of thick vegetation, but close to a shrub, such as *Haloxylon* (Paillat 1987; Gaucher 1987). Symens (1988) and Gaucher (1988) noted large numbers of ant colonies around houbara nests, which may provide an easily accessible food source for females and their broods.

Very little information exists on laying, incubation, hatching and chick-care by wild houbara bustards. Clutch size is usually two or three eggs, but occasionally one- or five-egg-clutches are laid. Only one clutch is produced per season, but replacement clutches can be laid if the first clutch is destroyed. During exceptionally dry years houbara may not breed at all.

Chicks are precocial and nidifugous at hatching, and are

brooded for the first 24 hours by the female (Plate 2.6). Live food is offered to the chicks, bill-to-bill, accompanied by soft vocalizations from the female. At 2 or 3 days of age, females drop prey on to the ground to be retrieved by the chicks. When 5 or 6 days' old the chicks start to feed independently. After 11 days the wings and shoulder areas of the chicks are covered by contour feathers, and at 4 weeks the birds are fully feathered, although the wings and tail are still quite short. They start flying short distances at the age of 1 month, but remain close to their mother for at least the next two months.

## 2.10 Status and Conservation

### 2.10.1 *Habitat Degradation and Over-hunting*

Populations of houbara are declining significantly in at least 15 of the 20 countries in its range: in his summary of the world status of the houbara, Collar (1980) listed excessive hunting (eight countries), overgrazing (four countries), agricultural development (two countries), general effects of civilization (one country) and egg collecting (in the Canary Islands) as probable reasons for this decline. The two major reasons are degradation of habitats through over-exploitation by man, and excessive hunting. With improved technology and modernization of agriculture, man has been able to intrude into formerly uninhabited semi-desert, desert and steppe areas. The construction of deep wells in arid regions, and the availability of water trucks to bedouin allows herdsmen to move into areas which previously were avoided due to lack of water. In many arid and semi-arid regions the numbers of sheep, goats and camels are increasing, greatly exceeding the carrying capacity of these ecologically sensitive habitats, with the result that complete destruction of the natural vegetation may occur. Livestock also comprise a disturbance that results in frequent nest desertion. In Israel, where hunting of houbara is banned, disturbance by livestock is one of the main factors preventing the houbara population from increasing (Lavee 1988). As a result of rapid intensification of agricultural activities, traditional breeding grounds are destroyed. Land is ploughed up and put down to seed, and natural plant communities are replaced by monocultures which are generally unsuitable for breeding birds.

During the last two decades hunting of houbara has increased, due in part to the increased availability of fire-arms, but mainly increased pressure from Arab falconry parties. Only a few decades ago Arab falconry was a quiet and sporting way of hunting, having little impact on houbara populations. However, with an increase in wealth following the oil boom, and with the availability of four-wheel-drive vehicles, which allow access to virtually every square kilometre of desert, large hunting parties are now common, equipped with dozens of cars, instruments for radio communication and navigation, and accompanied by local guides who are paid well for locating houbara. With several dozen trained falcons, large numbers of houbara may be killed by a single hunting party, exerting significant pressure on local populations (Haddane 1985; Mian 1984). For example, in Baluchistan in the winter of 1982–1983, hunting parties each killed several hundreds of houbara over a period of one-to-two months, and a progressive increase in the number of hunting parties visiting the region was reported (Mian 1984). The financial strength of most Arab countries and the dependence of many third-world countries on support from Arabia means that hunting parties can obtain permission to hunt in almost every country where houbara appear in significant numbers, even in areas where they are protected by law.

## 2.10.2 Declining Throughout the Range

The single most important breeding region was the **former USSR**, but between 1956 and 1979 a decline of 75 per cent was estimated, attributed to destruction of breeding habitat as a result of land improvement schemes and more intensive grazing, an increase in disturbance, local poaching, and large-scale hunting on wintering grounds (Alekseev 1985; Ponomareva 1985). In many areas only very small scattered and isolated populations have survived. An exception may be Kazakhstan, where large breeding populations persist (B. Gubin; O. Pereladova, pers. comm.). In **Pakistan**, a decrease in numbers of wintering birds has been noted (Malik 1985; Mian & Dasti 1985); in Cholistan this was estimated to be around 30 per cent (Mirza 1985). Although few data have been collected from **India, Iraq, Iran, Syria, Jordan, Egypt, Libya, Tunisia, Algeria,** and **Morocco**, populations have generally declined (Collar 1980; Clarke 1982; Haddane 1985; de Smet 1989; Razdan & Mansoori 1989; Saleh

1989), except in **Israel** where numbers appear to have remained constant (Mendelssohn 1980; Lavee 1988).

The small population of *fuertaventurae* on the **Canary Islands**, mainly on Fuerteventura, has been reduced in the second half of this century as a result of egg-collecting, hunting by local inhabitants, and in recent times by disturbance through an increasing number of visitors. Recent censuses suggested a total population of around 300 birds on Fuerteventura.

Little or no current information on breeding is available from the **Gulf States, Oman** and **Yemen**, where the species seems to have disappeared or to breed only very rarely (Gallagher & Woodcock 1980; Platt 1985; Osborne 1992). In **Saudi Arabia**, the only known breeding population is within the Harrat al-Harrah reserve in northern Saudi Arabia, where houbara are found in very low densities, and fewer than ten nests have been reported over the seven years since reserve creation (Seddon & van Heezik 1994). In the past, the houbara was not considered rare in Arabia and was found in all semi-desert areas on the peninsula. During winter, it was hunted irrespective of population density, and in spring bedouin collected eggs to supplement their diet.

### 2.10.3 Conservation Action

If the current trend continues, houbara populations may eventually collapse over most of the range of the species, reaching such low levels that recovery would be difficult. Such a decline would certainly mean houbara would no longer be able to be harvested by falconry, and this ancient Arab tradition would die. If the disappearance of houbara from many countries is to be avoided, then conservation action must be taken immediately to ensure the protection of habitats within the houbara's range, and to reduce the loss of animals from hunting.

Large-scale habitat conservation programmes are among the most promising steps for houbara conservation, particularly if hunting is prohibited or strictly controlled in such areas. Unfortunately, large-scale habitat conservation is very expensive and needs a stable political environment to be successfully implemented. In many countries houbara are legally protected, but laws are often not adequately enforced. Hunting provides economic input for some third-world countries, such as Pakistan, Tunisia and Algeria, and falconers are able to obtain permission for hunting, even in areas which otherwise are protected. Third-world countries often do not have the financial means efficiently to control and enforce legal conservation. Only with financial support from outside, or if governments use incoming funds from falconry parties to establish effective sustained yield management of their houbara populations, will such countries be able to protect the houbara successfully in the long-term.

An important approach to improving the situation is environmental education in Arab countries. It must be made known to hunters that houbara hunting can only be guaranteed for future generations if drastic bag limits are implemented now, and if hunting is discontinued in certain areas. Once Arabian hunters recognize their responsibility for ensuring the survival of the houbara, and see themselves as major representatives and guards for houbara conservation, the situation will certainly improve.

The central Asian breeding population of houbara is migratory and therefore can only be protected through international co-operation. International legal instruments, such as the Conservation of Migratory Species of Wild Animals (the 'Bonn Convention'), are important tools for the protection of such birds. However, representatives of countries 'utilizing' houbara, e.g. from Saudi Arabia and the Gulf States, need to become involved in the development of management strategies undertaken under such agreements. Co-ordination centres for the implementation of international conservation measures should be located in Arabian countries.

In view of our limited knowledge of the ecology, population biology and conservation status of the houbara, scientific research into these aspects has become a subject of major importance. Saudi Arabia and Abu Dhabi are both countries undertaking houbara research throughout the Arabian region. Recently captive breeding programmes have been established in various countries (Saudi Arabia, Abu Dhabi, Morocco, Uzbekistan), as a tool to ensure the long-term survival of the houbara.

Plate 2.1 The Houbara Bustard

Plate 2.2  Chlamydotis undulata undulata

Plate 2.3 Chlamydotis undulata macqueenii

Plate 2.4 Typical Habitat of Houbara Bustards

Plate 2.5  Houbara Bustard Nest

Plate 2.6  Houbara Chick

Plate 3.1 Aerial View of the Houbara Breeding Unit

Plate 3.2 A Block of Cages in the Breeding Unit

Plate 3.3 Inside a Cage in the Breeding Unit

Plate 4.1 Displaying Male Houbara

Plate 5.1 Female Houbara Passing Through a Size-selection Door

Plate 5.2 Pre-copulatory Display Directed Towards a Dummy Female

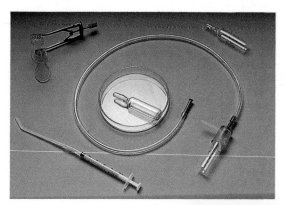

Plate 5.3 Semen Collection and Artificial Insemination Equipment: Insemination Syringe, Aspiration Apparatus, Diluent, Dish, and Forceps Used to Open the Cloaca

Plate 5.4 Collecting Equipment Used to Aspirate the Semen from the Cloaca into a Vial Held in the Operator's Hand

Plate 5.5 Insemination Procedure

Plate 6.1 The Incubation Room

Plate 7.1 Female Feeding a Chick

Plate 7.2 Plexiglas Brooding Boxes Containing Young Chicks

**Plate 7.3** Hand-feeding a Cricket to a Chick

**Plate 7.4** Hand-feeding Chicks

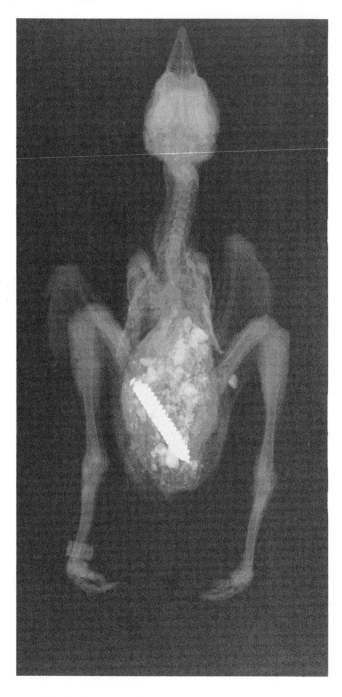

Plate 8.1 Presence of a Foreign Body (Screw) in the Ventriculus of a 1-week-old Chick (X-ray)

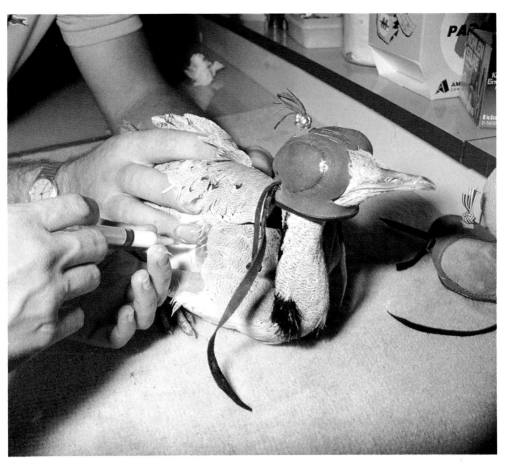
Plate 8.2 Vaccination of an Adult Houbara

Plate 9.1 Houbara Chicks

Plate 10.1 One-week-old Houbara Chick

Plate 10.2  Released Houbara in Flight in Mahazat as-Sayd

Plate 10.3 Houbara on Nest in Harrat al-Harrah

**Plate 10.4 Houbara Chick in the Wild**

# 3
# Management and Facilities

P. PAILLAT and P. GAUCHER

## 3.1 Composition of the Breeding Flock

Houbara caught as adults are easy to maintain in captivity, but remain very sensitive to stress, and in most cases fail to breed unless they are housed, undisturbed, in very large cages (Ramadan 1989). Since hand-reared houbara are most suitable for captive breeding (Mendelssohn et al. 1983), we decided that the breeding flock should mostly comprise hand-reared birds. For this purpose, five expeditions were undertaken to different houbara breeding areas: three to Pakistan to collect eggs and chicks of the Asiatic sub-species (*Chlamydotis undulata macqueenii*), suitable for future reintroduction in Saudi Arabia, and two to Algeria to collect from the north African sub-species (*Chlamydotis undulata undulata*, hereafter referred to as *macqueenii* and *undulata* respectively). The latter were to be used as experimental birds to improve captive management.

### 3.1.1 Egg Collection

The Algerian expeditions (1986 and 1987) took place in Ouled Djellal, south of Biskra, and in the Abiod Sidi Le Sheikh region. Expeditions to Pakistan in 1987 and 1988 visited the districts south and south-west of Kharan, Baluchistan, and the Chagai Plain of Waziristan.

Each expedition was equipped with forced-air incubators, hatchers and brooders (manufacturer: Grumbach). Eggs were transported from the nest to the camp in a protective block of foam. They were then candled to estimate date of hatching, and warm eggs were directly transferred into incubators, while eggs found cold were pre-heated for two hours at about 30°C before being

### Table 3.1 Origin of Birds in the NWRC Breeding Flock (January 1994)

| Origin | C. u. macqueenii | | C. u. undulata | | Hybrid | |
|---|---|---|---|---|---|---|
| | male | female | male | female | male | female |
| Wild Pakistan | 10 | 14 | | | | |
| Wild KSA | 2 | 4 | | | | |
| Jordan | 2 | 2 | | | | |
| Pakistan 1986 | 3 | 3 | | | | |
| Pakistan 1987 | 14 | 13 | | | | |
| Pakistan 1988 | 9 | 9 | | | | |
| Algeria 1986 | | | 9 | 13 | | |
| Algeria 1987 | | | 18 | 20 | | |
| NWRC 1989 | 1 | 3 | 1 | 1 | | |
| NWRC 1990 | 1 | 3 | 8 | 7 | | |
| NWRC 1991 | 11 | 10 | 7 | 6 | | |
| NWRC 1992 | 17 | 26 | 14 | 14 | 2 | 2 |
| NWRC 1993 | 34 | 35 | 35 | 49 | | |
| Total Founders | 40 | 45 | 27 | 33 | | |
| Total Born NWRC | 64 | 77 | 65 | 77 | 2 | 2 |
| **Total** | 104 | 122 | 92 | 110 | 2 | 2 |

placed in the incubator. Incubation temperature was maintained at 37.5°C and humidity at 45 per cent. Eggs were candled every third day; when the shape of the air chamber indicated imminent pipping, they were transferred to the hatcher. After hatching, chicks' navels were disinfected, they were weighed, and transferred into a brooder. In the hatchers the temperature was lower (37°C) and humidity higher (60–70 per cent). Brooders were fitted with a heater providing a temperature of 37°C in one corner. The first meal was given 24 hours after hatching. A total of 93 chicks were collected during the spring expeditions in Pakistan, and an extra 20 juvenile birds were collected in autumn 1988.

In Algeria, 207 eggs were collected, from which 175 chicks hatched successfully. Twelve of the hatchlings were given to the Algerian government and 34 died before being shipped to Saudi Arabia, mainly because of transportation at too early an age, and inadequate food.

### 3.1.2 Current Bird Stock

The initial breeding flock held at the NWRC consisted of the above birds as well as some donated from private collections.

### Table 3.2 Age of the Breeding Flock Between 1988 and 1993

**C. u. macqueenii**

| Age | 1988 | 1989 | 1990 | 1991 | 1992 | 1993 |
|---|---|---|---|---|---|---|
| 1 yr males | 23 | 12 | 3 | 6 | 12 | 23 |
| females | 23 | 14 | 3 | 6 | 11 | 26 |
| 2 yr males | 14 | 23 | 12 | 1 | 5 | 11 |
| females | 10 | 23 | 14 | 3 | 6 | 11 |
| 3 yr males |  | 11 | 18 | 12 | 1 | 4 |
| females |  | 10 | 22 | 14 | 3 | 6 |
| 4 yr males |  |  | 9 | 18 | 11 | 1 |
| females |  |  | 9 | 20 | 11 | 3 |
| 5 yr males |  |  |  | 8 | 16 | 10 |
| females |  |  |  | 8 | 17 | 11 |
| >5 yr males | 3 | 2 | 2 | 2 | 8 | 23 |
| females | 3 | 2 | 2 | 2 | 9 | 24 |

**C. u. undulata**

| Age | 1988 | 1989 | 1990 | 1991 | 1992 | 1993 |
|---|---|---|---|---|---|---|
| 1 yr males | 30 | 0 | 2 | 10 | 9 | 15 |
| females | 37 | 0 | 1 | 9 | 7 | 17 |
| 2 yr males | 18 | 30 | 0 | 1 | 9 | 8 |
| females | 23 | 37 | 0 | 1 | 8 | 7 |
| 3 yr males |  |  | 13 | 23 | 0 | 1 | 9 |
| females |  |  | 23 | 32 | 0 | 1 | 7 |
| 4 yr males |  |  |  | 11 | 21 | 0 | 1 |
| females |  |  |  | 18 | 26 | 0 | 1 |
| 5 yr males |  |  |  |  | 10 | 21 | 0 |
| females |  |  |  |  | 17 | 22 | 0 |
| >5 yr males |  | 2 | 1 | 0 | 0 | 8 | 27 |
| females |  | 2 | 1 | 0 | 0 | 14 | 36 |

Three management strategies were attempted to increase the size of the breeding flock: (1) birds were isolated for artificial insemination; (2) birds were kept in pairs; and (3) birds were kept in heterosexual groups. By January 1994, the flock comprised 432 birds, whose origins are indicated in Table 3.1. Ages of birds and management of the flock between 1988 and 1993 are given in Tables 3.2 and 3.3.

## 3.2 Facilities

### 3.2.1 *Breeding Cages*

The cages used within the breeding unit are laid out in blocks 40 m. × 12 m.) of 20 (Fig. 3.1; Plate 3.1). Cages are arranged in

### Table 3.3 Management of the Breeding Flock Between 1988 and 1993
(Birds Aged Two Years and Over)

*C. u. macqueenii*

|  | 1988 | 1989 | 1990 | 1991 | 1992 | 1993 |
|---|---|---|---|---|---|---|
| Lone males | 0 | 2 | 38 | 29 | 38 | 46 |
| females | 0 | 0 | 33 | 31 | 42 | 50 |
| Paired males | 11 | 29 | 1 | 11 | 0 | 0 |
| females | 10 | 29 | 1 | 11 | 0 | 0 |
| Grouped males | 6 | 11 | 5 | 1 | 3 | 3 |
| females | 3 | 14 | 13 | 2 | 4 | 4 |

*C. u. undulata*

|  | 1988 | 1989 | 1990 | 1991 | 1992 | 1993 |
|---|---|---|---|---|---|---|
| Lone males | 0 | 1 | 18 | 12 | 34 | 45 |
| females | 0 | 15 | 30 | 14 | 40 | 51 |
| Paired males | 14 | 32 | 0 | 0 | 0 | 0 |
| females | 14 | 32 | 0 | 0 | 0 | 0 |
| Grouped males | 6 | 11 | 16 | 20 | 5 | 0 |
| females | 11 | 14 | 20 | 30 | 5 | 0 |

two lines of 10, back-to-back (Plate 3.2). Each cage (6 m. × 4 m. × 2.3 m.) consists of an 80 cm.-high block wall, on top of which is fixed a tubular metal frame covered with a 5 cm. × 5 cm. square mesh. A 1 m.-wide mesh door at the front of the cage is sometimes fitted with a wooden screen on the lower part of the door, to prevent birds being disturbed by seeing others in adjacent blocks, or staff in the walkways (Plate 3.3).

The side- and back-walls of each cage have an opening (approx. 1.7 m. × 0.5 m.) leading into adjacent cages. Wooden screens placed over these openings enable us to control the number of cages that birds have access to. A mesh panel can be substituted to allow the birds to have visual contact with other birds, but no physical contact.

Each cage has an overhead panel (1.22 m. 1.22 m.) providing shelter from heavy rain and strong sun, and an alfalfa plot (3.7 m. × 1.7 m.) at the back of the cage, watered from an overhead sprinkler system controlled from switches outside the cage (Fig. 3.2). The remainder of the floor area is covered with clean sand, and cages containing females have a few large stones placed to provide nesting cover.

The breeding unit consists of 20 of these blocks laid out on a five-by-four block grid, with 4 m.-wide walkways between each block. The interior floor level of the cages is 30 cm. above this

*Management and Facilities*

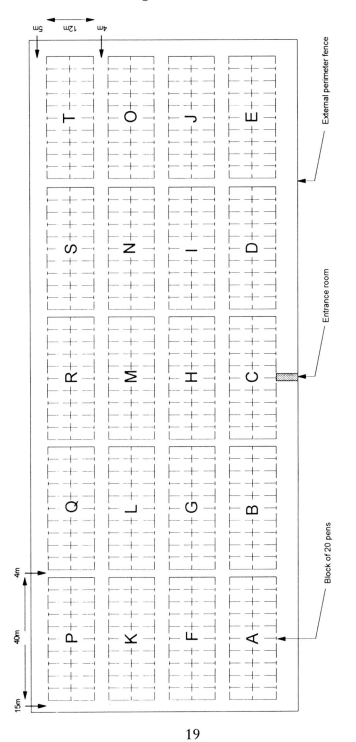

**Figure 3.1 Layout of houbara breeding pens**

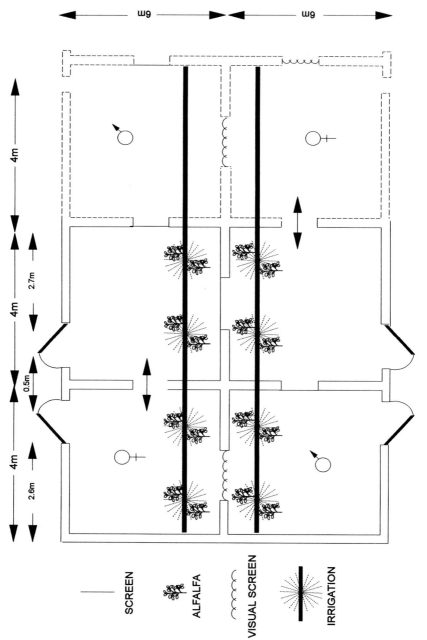

**Figure 3.2** Layout of part of a block within the houbara breeding unit

walkway. The entire unit is surrounded by a frame supporting a 1.5 cm. square mesh to prevent small birds from entering the unit and introducing diseases. The outer cage has a large access door to enable vehicles to enter the unit for cleaning and maintenance when the houbara are in the wintering cages. Otherwise access to the unit is via an entrance/changing room, where staff put on overalls and rubber boots before entering the breeding unit through a disinfectant foot dip.

A 2 m.-high predator-proof fence around the breeding unit protects breeding birds from foxes and cats. A video monitoring system was installed in one block to observe and record the displays and behaviour of the birds.

### 3.2.2 Wintering Cages

Birds are moved to the wintering cages at the end of the breeding season to allow cleaning of the breeding cages. These consist of nine pens (30 m. × 10 m.) constructed from a fine mesh fitted over a tubular metal frame. Soft plastic mesh 1.5 m. high lines the lower, inner side of these cages to prevent birds from injuring themselves. Pens are laid out in a block of six double cages in a line of three-by-two and one double cage to the side. The outer cages have a thick canvas wind-break fitted to a height of 1.5 m. along their shortest length, to protect the birds from strong winds and sand storms.

### 3.2.3 Incubation Facilities

The incubation complex (84 $m^2$) comprises an entrance/egg-cleaning room, egg-storage room, fumigation room, air-conditioning room, incubation room, hatching room and monitoring office (Fig. 3.3). The internal materials were chosen to facilitate hygiene: ceramic tiles, marble, epoxy paint and Plexiglas.

A positive pressure state is maintained in the egg-cleaning room, ensuring an outflow of air when the entrance door is opened. A footbath is located in front of the internal door. In the egg-storage room, temperature and humidity are controlled independently of the rest of the unit, by controls situated outside the room.

The fumigation room is fitted with a fan through the outside wall to allow extraction of gases when the room is in operation; this opening is protected on the outside by a steel canopy which contains a disposable synthetic filter, to prevent entry of dust

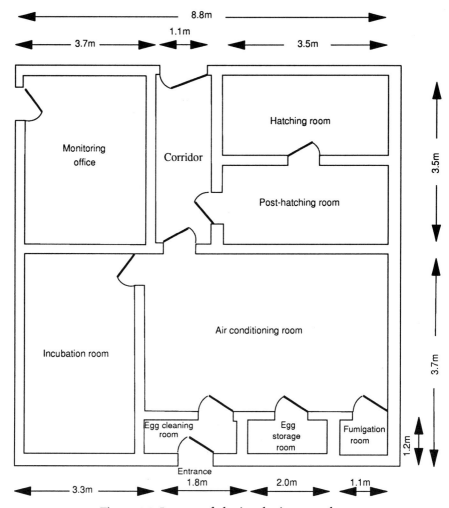

Figure 3.3 Layout of the incubation complex

when the fan is not in use. Two ventilators at the bottom of the door are closed during fumigation and opened when the gas is extracted.

The air-conditioning room is the only means of access to the egg-storage room, fumigation room and incubation room. All the air used within the complex, with the exception of the hatching room and monitoring office, enters this room via external ducting. A plastic pipe at the bottom of the ducting is connected to a pressure pump, and sprays a solution of filtered water and disinfectant (with both anti-bacterial and virucidal activity) upwards into the ducting (Fig. 3.4). Disinfectant is renewed every 24 hours. A metal container placed beneath the ducting collects and diverts excess water.

On the internal side of the air-opening Plexiglas ducting contains two filtration points; one immediately as the air enters the room and a second 130 cm. from the first. Each point contains a disposable synthetic filter which has an 86 per cent arrestance specification. The end of the ducting has a 90 degree bend, directed upwards towards the air-conditioning unit, which permits the incoming air to be heated or cooled as necessary. A large stainless steel sink unit is fitted beneath the Plexiglas ducting to allow washing and fumigation of all equipment within the complex.

Exit from the complex and access to other rooms is through a steel door connecting to a corridor leading to the external door; this is fitted with an extraction fan 30 cm. from the top, maintaining a state of slight positive pressure in the corridor, which prevents entry of air.

Entry to the incubation room is via the air-conditioning room. The incubation room (Fig. 3.5) is designed to operate in a positive pressure condition, and therefore it is necessary to control the expulsion of air through outlets to maintain a balance between the number of air changes required and the desired degree of positive pressure. This is achieved in two ways. First, adjustable louvres are fitted to each outlet flush with the room wall to enable the control of air outflow. Secondly, the external side of the outlet is fitted with a galvanized steel canopy fitted with a sponge filter 5 cm. thick. The latter is designed to prevent any back pressure created by climatic conditions (i.e. wind). Positive pressure is created by the use of three fans placed in the dividing wall between the incubator room and the air-conditioning room. These are controlled independently by variable reduction switches, so that a balance can be achieved by controlling the speed of each fan and the adjustment of the outlets. The small size of the incubation room relative to that of the air-conditioning room means that it is possible to achieve minimal variations in temperature. Temperature, humidity and $CO_2$ levels are monitored within the incubators from a data logger in the monitoring office.

Hatching and post-hatching rooms have a separate air source from the rest of the complex, because they are likely to be subjected to the highest bacterial load. The monitoring office is entered directly from the outside of the complex; incubation can

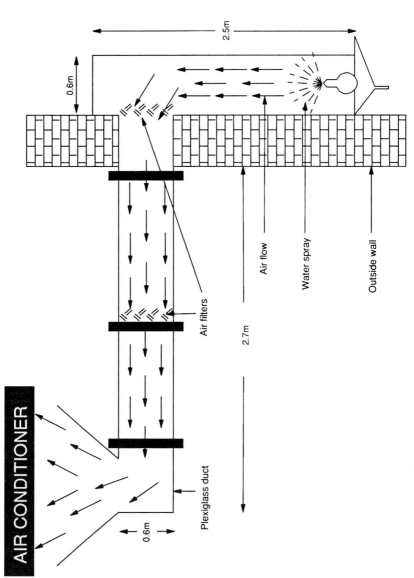

**Figure 3.4 Layout of air intake to the air conditioning room**

## Management and Facilities

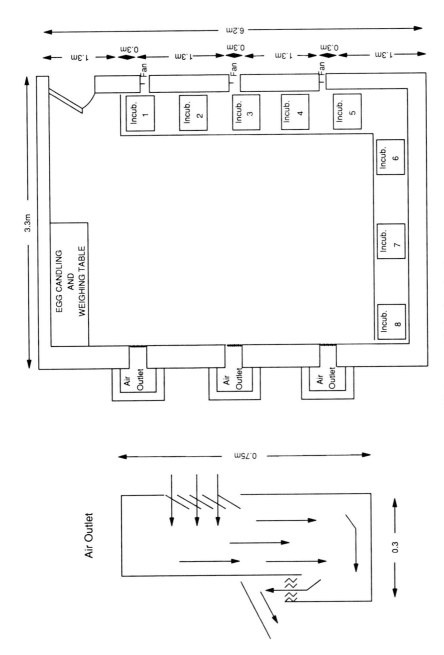

**Figure 3.5 Layout of the incubation room**

be monitored from this room without compromising hygiene through repeated entry.

### 3.2.4 Rearing Unit

The rearing unit (Fig. 3.6) was constructed to allow flexibility in rearing regimes for chicks reared for different purposes; i.e. as breeding stock, or for release. Internal materials used allow us to maintain a high level of hygiene. Food preparation tables have glass work-surfaces. The room designed for the initial rearing stage has sealed and painted cement rearing tables, which contain the sterilized sand that is used as the rearing medium.

After 15 days chicks are transferred into rooms with a sand floor that are connected to external pens through a hole in the wall. A low lamp is provided for warmth. The external pen has soft walls made from tensioned shade-cloth.

### 3.2.5 Quarantine Facilities

The quarantine facility comprises ten separate cages, covered with sparrow-proof netting. Four of these cages have a concrete floor to allow easy disinfection, and six have earthen floors, which are more comfortable for the birds. These are reserved for non-contagious or injured birds, and for secondary quarantine. Special overalls and boots are used when working in the quarantine facility.

## 3.3 Management of Birds

### 3.3.1 Handling

In any captive breeding project, one of the major problems encountered is catching and handling the breeding species. When artificial insemination techniques need to be applied, the risk factor is greatly increased. The method of handling described below is not meant to be an authoritative guide, but illustrates how we have significantly reduced injuries.

The legs and wings of houbara are extremely fragile, and feathers tear out easily. Only one person should attempt to catch the bird in its cage. A circular landing net approximately 70 cm. in diameter and 70 cm. in depth, with a handle length of 2 m., and covered with opaque black cloth, should be placed on top

## Management and Facilities

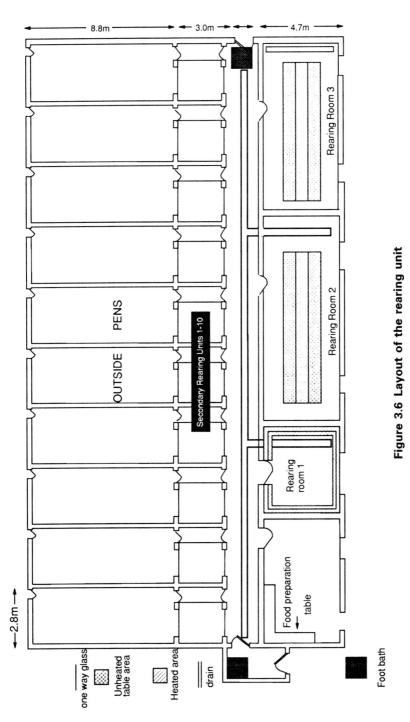

**Figure 3.6 Layout of the rearing unit**

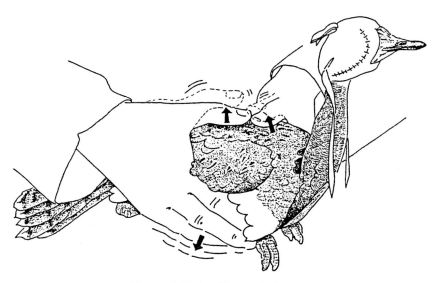

Figure 3.7 Handling technique

of the bird; the darkness causes it to quieten. The bird should be grasped under the bag by maintaining its wings close to the body with both hands, the thumbs lying on its back with the other fingers wrapping around the wings and the body just below the elbow joint. The bird is then raised quickly from the ground to prevent the risk of leg fracture or slipped tendon caused when the bird exerts pressure on the ground in an attempt to escape. The head is immediately covered with a falcon hood, which calms the bird instantly; it should then be returned to the ground and made to assume a resting position by gently exerting pressure on its back (Fig. 3.7).

When lifted, the index and second fingers on both hands should also hold its legs close to its body, allowing some movement and avoiding exerting too much pressure on the legs. The bird can now be weighed, vaccinated, inseminated, etc. without major problems. When released, the bird should be placed on the ground in a resting position and the hood removed. The birds in the breeding unit are either feather-cut or pinioned to avoid possible damage to themselves when disturbed.

*Management and Facilities*

## 3.3.2 Annual Schedule and Sanitation

During the breeding season faeces and feathers are removed each week from all cages. At the end of each breeding season all cages in the breeding unit are emptied for cleaning. All birds are caught, vaccinated, weighed, dewormed and moved to the winter pens.

In the breeding cages all debris, feathers, droppings, dead alfalfa, etc. are removed from cages and walkways. The cage floors are washed down with a 1 per cent solution of caustic soda, and after two days sprayed with an oocide to kill eggs of coccidia and tapeworm. Any routine maintenance work is carried out at this time, after which the cages are given a thorough wash with a pressure washer, removing ingrained dirt from cage walls and mesh. Both the cages and the walkways are then sprayed from top to bottom with a 2 per cent solution of formalin.

A 50–cm. area around each block wall is sprayed with a long-acting insecticide, to form a protective barrier against ants and beetles, which are intermediate hosts in the tapeworm's life-cycle. This spraying is repeated every six weeks to ensure that the barrier is maintained. Finally, the alfalfa plots are replanted where necessary, and the pens are left vacant for a period of four months.

The birds are returned to the breeding unit at about the end of December. Once again, all the birds are vaccinated, dewormed, weighed and are allocated to one of two breeding systems: the birds used for artificial insemination are kept isolated, but with visual contact between males and females in some cases, while the birds used for 'natural mating' are housed together.

In the incubation complex, disinfection is particularly important, and must be done carefully and thoroughly. At the end of the breeding season incubators are cleaned of all organic matter using vacuum cleaners, and they are then left empty for three months. One month before the start of the new season, they are cleaned again and disinfected with ammonium IV. Finally, two weeks before being used, they are sprayed twice with a formalin solution, with a two-day interval between sprayings. It is also important to clean thoroughly all cables and pipes connected to the incubators. All other equipment and movable furnishings are cleaned at the same time, and the positive pressure system filters are changed.

Rooms are cleaned and disinfected at the end of the breeding season, and again three months later. Before equipment is replaced, the rooms are also sprayed with a formalin solution. While in operation it is important that the positive pressure system is functioning. Bacteriological examination of the incubators, rooms and equipment is highly recommended. Many bacteriological counting media (kits) are available; these can easily be used before and after disinfection, to check how effective the disinfection has been. The results from such kits are immediate, rather than available only after laboratory analysis. Opening of doors should be kept to a minimum while the complex is in operation, to prevent dust from entering the unit.

In the rearing unit, at the end of each breeding season all movable objects are taken out of the building, cleaned and disinfected with a solution of ammonium IV, and stored. Before being replaced just before rearing resumes, they are cleaned and disinfected again.

All walls, floors, ceilings and benches in the rearing unit are washed initially with water, followed by disinfection with ammonium IV; this is carried out twice, with an interval of 15 days between cleanings. The unit is then quarantined for a period of at least two months. As the breeding season approaches, the unit is disinfected again with ammonium IV, between two and four times, 15 days apart. Finally, a solution of formalin is sprayed on all surfaces; it is necessary to ventilate well afterwards. Cleaning agents containing chlorine should be avoided, as solutions may evaporate on hot days and cause slight problems in the upper respiratory tracts of the chicks.

Whilst in operation, staff entering the unit must wear special overalls and boots, and enter through a footbath containing a Chlorox solution.

### 3.3.3 General Sanitary Measures

Vehicles entering the fenced NWRC compound must drive through a dip containing a formalin-based disinfectant, which reduces the risk of vehicles bringing in dirt containing pathogens on their tyres.

All staff must put on overalls and rubber boots before entering the breeding unit, quarantine or wintering cages. Different clothing is used in each facility. Staff must also wash their hands with an anti-bacterial soap before handling any of the birds, and must pass through a chlorine-based disinfectant foot dip before entering the unit itself.

## Management and Facilities

Any new birds arriving at the Center, or birds being returned to the unit from field experiments, are placed in quarantine to allow their health status to be monitored before being introduced into the unit. The quarantine facility is situated well away from the breeding unit to ensure that contamination does not reach the unit. Food and water dishes are cleaned weekly using ammonium IV disinfectant.

### 3.3.4 Identification

Immediately upon hatching, each chick is fitted with an expandable numbered plastic ring. This is worn until four months of age, at which time the leg is large enough to wear a permanent metal identification ring. The metal rings are uniquely numbered split rings.

### 3.3.5 Food

The composition of the artificial diet fed to any captive species is one of the main factors determining reproductive success. Wild houbara appear to be opportunistic and omnivorous (see Chapter 2; section 2.6). After some experimentation with different combinations of animal and vegetable food, we settled on a diet comprising pasteurised pellets; those fed to birds in winter contain 14 per cent protein, whereas those fed during the breeding season contain 22 per cent protein. Females are also provided with powdered oyster shell, and all cages contain fresh alfalfa. A polyvitamin complex is added to the drinking water twice a week.

Between December 1991 and January 1993, food consumption was measured every day. We observed variations in food intake during the annual cycle: consumption was least in May, at the end of the reproductive season, and in autumn, at the end of the moult. Food intake was greatest at the beginning of the breeding cycle, and in the middle of the moult. Mean maximum consumption was 45g./day for females and 60g./day for males; mean minimum intake was 30g./day for females and 43g./day for males.

# 4

# Reproductive Parameters in Captivity

M. SAINT JALME, P. GAUCHER and P. PAILLAT

## 4.1 Introduction

In the following chapter information on reproductive parameters of captive houbara bustards is presented. The reproductive cycle of wild houbara is still largely unknown. Timing of nesting appears to be quite variable across the species' breeding range, but is concentrated in the spring months. In North Africa eggs are usually laid in March-April (Cramp & Simmons 1980), although according to Heim de Balzac and Mayaud (1962) egg-laying can occur between January and June. In Syria eggs are found from March to May (Cramp & Simmons 1980) and in the southern former USSR from April to June (Dement'ev & Gladkov 1968). Occasionally, eggs have been found in Algeria as early as November, or as late as June (Cramp & Simmons 1980).

Survival of a species in a semi-arid environment requires individuals to adjust to changing environmental conditions, and also to reproduce at a time of year which will be favourable for survival of young. The annual photoperiodic cycle is well-known as the most significant environmental variable, synchronizing seasonal reproductive activities in species breeding at middle and high latitudes where seasons are temporally well defined (Farner & Gwinner 1980), but little is known about which parameters predict suitable environmental conditions in arid climates, and control the timing of reproduction. Temperature, precipitation and food availability are certainly of utmost importance. For houbara, some authors have suggested that rainfall is important to trigger reproduction: breeding does not occur during dry years in Africa (Etchécopar & Hüe 1978) and in Israel most houbara do not breed after winters with little or no rain (Mendelssohn 1980). In Saudi Arabia, the occurrence of rain is relatively predictable, with a high probability of rainfall between March and April (Abouammoh 1991). Eggs laid in February will hatch in March, when rainfall and the subsequent increase in food availability are most likely to occur.

Age of first breeding of houbara is not known (Cramp & Simmons 1980), although sexual maturity is thought to be reached after two years (Mendelssohn 1983). These parameters provide further demographic information: for example, deferred maturity is generally found in long-lived species, where the loss of a few breeding seasons has less impact on life-time reproductive potential than in short-lived species. In such species with low annual mortality, clutch size is often small and foraging techniques may take some time to perfect (Hall et al. 1987). Houbara bustards are reported to produce only one clutch of one-to-four eggs, laid on alternate days (Heim de Balzac & Mayaud 1962; Etchécopar & Hüe 1978; Mendelssohn 1980; Urban et al. 1986). A replacement clutch may be laid if the first is lost (Cramp & Simmons 1980).

We examined the following parameters in our captive flock: (1) the annual cycle of laying; (2) the influence of environmental factors on reproduction; (3) age at first reproduction; (4) timing of laying; (5) laying and clutch interval; (6) number of clutches in a reproductive season; and (7) reproductive potential.

## 4.2 Methods

The composition of the breeding stock and management of the birds from 1990 onwards have been described in Chapter 3. Birds aged between 1 and 12 years were kept in the breeding unit during winter and spring, and in heterosexual groups of 15 to 20 birds during summer and autumn. Dates of transfer of birds to the breeding units in the years 1990 to 1993 were 21 January, 27 January, 20 December, and 11 November respectively.

Males were checked for display twice a day from the date of transfer until August (Plate 4.1). Females were checked four times a day to collect eggs. Most eggs were collected for artificial incubation just after laying; this procedure is known as egg-pulling. On some occasions, and generally at the end of the reproductive cycle, the eggs were left under the female for natural incubation in order to reduce imprinting of chicks on human beings, and thereby to produce birds for release experiments. To

### Table 4.1 Date of First Egg Laid

|                | 1990    | 1991    | 1992    | 1993    |
|----------------|---------|---------|---------|---------|
| C. u. macqueenii | 13 Feb. | 12 Feb. | 20 Jan. | 13 Dec. |
| C. u. undulata   | 22 Feb. | 24 Feb. | 29 Jan. | 6 Jan.  |

analyse the degree of breeding synchrony among members of both sexes within each season, we have taken the laying date of the first egg for females and the date of first display for males.

A meteorological station was installed at the NWRC towards the end of 1989. Air temperature and rainfall were recorded systematically from 1990 onwards. In 1992, a study of the annual cycle of houbara kept under standard conditions was conducted on 24 *macqueenii* and 24 *undulata* (24 males and 24 females). All birds were kept isolated in individual pens, and food and water were provided *ad libitum*. Every 15 days body mass of birds and stage of moult were recorded. Eggs were collected every day, and displaying males were recorded.

## 4.3 Results

### 4.3.1 Annual Reproductive Cycle

Although we found seasonal variation in the pattern and timing of the onset of laying for both subspecies, probably due to annual variations in ambient temperature and rainfall (Fig. 4.1), houbara showed a seasonal breeding pattern (see Table 4.1). In each year laying generally begins after maximum ambient temperatures decrease to below 20°C, and stops when maximum temperatures exceed 30°C (Fig. 4.1). Recommencement of laying later in the season, and a resurgence in the occurrence of male displays may follow significant rainfall, which tends to cause a temporary decrease in ambient temperatures. In these cases, laying occurs 15 days after rainfall. Since males were observed displaying as early as October, and one female laid in December when photophase was still decreasing, photoperiod does not seem to have an impact on the onset of the reproductive season. Instead, rainfall could act as a secondary 'Zeitgeber', and high ambient temperatures could be the major factor terminating the breeding season.

Peaks of laying are synchronous; the maximum number of eggs was always laid in the middle of March. Females generally

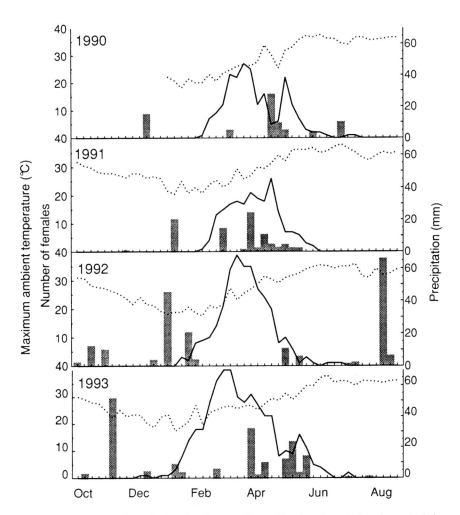

Figure 4.1 Seasonal variation in the number of laying female houbara (solid line), precipitation (histograms) and maximum ambient temperatures (dotted line): 1990–1993.

stopped laying at the beginning of June, except for one or two females which laid a clutch in July.

Annual variations in body mass, number of displaying males, and rate of moult (number of primaries simultaneously in growth) of 12 male *macqueenii* and 12 male *undulata* kept in standard conditions are shown in Fig. 4.2. Seasonal variation of these three parameters is identical in both sub-species. Males begin to display in December and stop displaying with the onset of moult, between April and August. In both subspecies, the highest percentage of males displaying was recorded at the beginning of March. Body mass peaks at the beginning of the breeding season, and then decreases slowly till the onset of moult. Displays begin when body mass reaches its maximum. Annual cycles of body mass and moult are similar in females (Fig. 4.3): laying begins when body mass peaks, and stops with the onset of moult, when body mass is at a minimum.

In both subspecies the reproductive season of males is longer than in females. Peaks in occurrence of displays are observed 15 to 30 days before the peak of laying, this lag being more pronounced in *undulata* than in *macqueenii*.

### 4.3.2 Reproductive Parameters

#### 4.3.2.1 Age at First Breeding

We recorded the age of sexual maturity in females (first egg laid) of both subspecies *macqueenii* and *undulata* between 1988 and 1993, for both founders and birds hatched in captivity at Taif (Table 4.2). Only 32 per cent of founder *macqueenii* and 15 per cent of *undulata* began to lay at 2 years of age. By 4 years of age, 89 per cent *macqueenii* and 69 per cent *undulata* had reached sexual maturity, and some females took as long as five or more years before beginning to lay. As yet we do not know if this is a natural reproductive strategy of the species, or if it is due to the management of our birds in captivity. We found no differences in age of sexual maturity between sub-species, or between founders and the first captive generation. In males 30 per cent to 40 per cent began displaying at the age of 1 year, 90 per cent after two years, and all males were displaying by 3 years of age. Only a small proportion of the first captive-bred generation of *macqueenii* (6.5 per cent) laid eggs at 1 year of age.

## Reproductive Parameters in Captivity

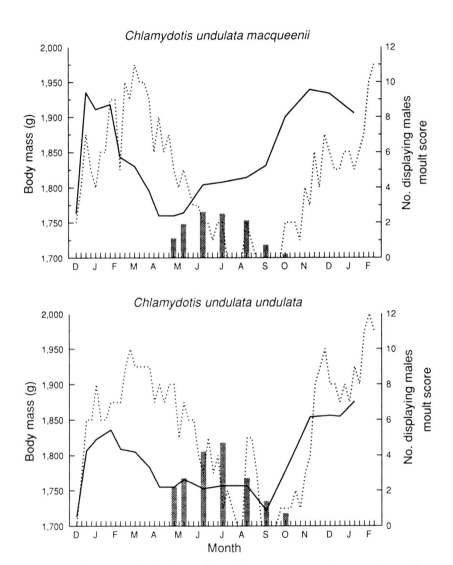

Figure 4.2 Annual variation in male houbara body mass (solid line), number of displaying males (dotted line) and rate of moult, i.e. number of primaries simultaneously in growth (histograms)

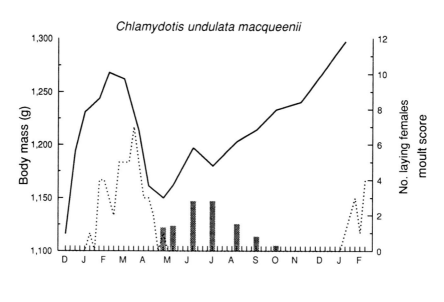
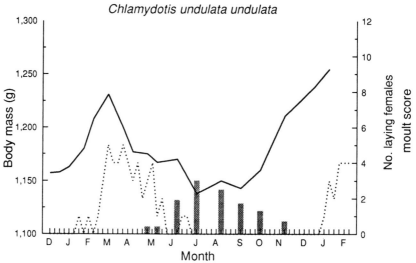

Figure 4.3 Annual variation in female houbara body mass (solid line), number of laying females (dotted line) and rate of moult, i.e. number of primaries simultaneously in growth (histograms)

## Reproductive Parameters in Captivity

**Table 4.2 Relationship between Age and Laying (Percentage of Laying Females in Five Age Classes)**

Founder

|  |  | 1 yr | 2 yr | 3 yr | 4 yr | >4 yr |
|---|---|---|---|---|---|---|
| C. u. | N | 47 | 47.0 | 47.0 | 47.0 | 47.0 |
| macqueenii | % | 0 | 31.9 | 72.3 | 89.4 | 91.5 |
| C. u. | N | 60 | 60 | 57 | 49.0 | 48.0 |
| undulata | % | 0 | 15 | 56.1 | 69.4 | 85.4 |

First generation captive-bred

|  |  | 1 yr | 2 yr | 3 yr | 4 yr | >4 yr |
|---|---|---|---|---|---|---|
| C. u. | N | 46.0 | 18.0 | 9.0 | 3 | 0 |
| macqueenii | % | 6.5 | 27.8 | 55.6 | 100 |  |
| C. u. | N | 30 | 16 | 9.0 | 0 | 0 |
| undulata | % | 0 | 25 | 44.4 |  |  |

*queenii* (6.5 per cent) laid eggs at 1 year of age.

### 4.3.2.2 Clutch Size, Inter-clutch Interval and Laying Capacity

Clutch size and inter-clutch interval were determined assuming that two eggs separated by less than five days belonged to the same clutch. Within a clutch, eggs were laid on alternate days 80 per cent of the time. The remaining 20 per cent were spread over three to four days. In such cases we considered stress or other environmental factors to be responsible for delayed oviposition or yolk resorption. As we found intra-individual variation in inter-clutch interval from the beginning to the end of the laying period of each female, in our analyses we used only the interval between the three first clutches. We also found intra-individual variation in clutch size between the first and the last clutch: therefore, to compare clutch size between sub-species and age classes, we only used the maximum clutch size of each female. The number of eggs laid per reproductive season gives an indication of the laying potential of female houbara in captivity. Given that a significant proportion of females incubated some of their own eggs, these results represent the minimum possible. Table 4.3 shows reproductive parameters of the two sub-species for the 1993 reproductive season. All females were isolated in individual cages.

The interval between clutches is greater in *undulata* (n = 60 intervals, mean ± SD = 11.2 ± 4.1 days) than in *macqueenii* (n = 61 intervals, mean = 9.3 ± 4.6 days), with no difference between age classes. The average maximum clutch size is larger in *mac-*

**Table 4.3 Percentage of Laying Females and Production of Eggs in Each Age Class in 1993**

*C.u. macqueenii*

| Age | 1 yr | 2 yr | 3 yr | >3 yr | Total |
|---|---|---|---|---|---|
| No. females | 27 | 10 | 6 | 38 | 81 |
| No. laying females | 3 (11%) | 4 (40%) | 3 (50%) | 29 (76%) | 39 (48%) |
| No. eggs | 5 | 16 | 20 | 341 | 382 |
| Mean max. clutch size | 1.7 (1–2) | 1.8 (1–4) | 3.0 (2–4) | 2.7 (1–4) | 2.6 (1–4) |
| Inter-clutch interval | | 7 (6–8) | 9 (6–12) | 10 (5–30) | 9 (5–30) |
| Mean no. eggs/female | 2 (1–2) | 4 (1–13) | 7 (3–10) | 12 (1–30) | 10 (1–30) |

*C.u. undulata*

| Age | 1 yr | 2 yr | 3 yr | >3 yr | Total |
|---|---|---|---|---|---|
| No. females | 17 | 7 | 7 | 37 | 68 |
| No. laying females | 0 | 2 (29%) | 3 (43%) | 31 (84%) | 36 (53%) |
| No. eggs | 0 | 8 | 6 | 274 | 288 |
| Mean max. clutch size | | 1.5 (1–2) | 1.7 (1–2) | 2.1 (1–3) | 2.1 (1–3) |
| Inter-clutch interval | | 11 (9–14) | | 11 (6–25) | 11 (6–25) |
| Mean no. eggs/female | | 4 (1–7) | 2 (1–3) | 9 (1–20) | 8 (1–20) |

*queenii* (n = 36 clutches, mean = 2.6 ± 1 eggs) than in *undulata* (n = 35 clutches, mean = 2.1 ± 0.7 eggs). In *macqueenii*, 68 per cent of the clutches comprise one or two eggs and 30 per cent three or four eggs. In *undulata* 80 per cent of clutches consist of one or two eggs. A clutch greater than three eggs was never recorded in *undulata*.

The laying capacity, i.e. the number of eggs laid by a female, does not differ between sub-species, but does increase with age. The mean number of eggs laid by *macqueenii* females increased from 1.7 at 1 year old, to 11.8 from birds aged more than 3 years (Table 4.3). Nevertheless, there is great variability in the number of eggs laid within each age class. The maximum number of eggs laid by a single female in one reproductive season was 30 for *macqueenii* and 20 for *undulata*.

Although there is considerable variation in the timing of the start of laying in both sub-species and in each age-class, older *macqueenii* females tend to lay earlier in the season, and 3– and 4–year-old birds lay earlier than 2–year-old birds. This delay in seasonal maturation explains in part why 2–year-olds laid fewer eggs per laying season than birds older than 3–years.

### 4.3.2.3 Number of Days of Display and Timing

While male display is not comparable to laying in females, it is a useful indicator of male sexual maturity. The number and timing of displays were recorded between December 1991 and July 1992. Dates of first displays were highly variable: the first *macqueenii* males commenced displaying as early as the middle of December, whereas the last began in the last week of May. In both subspecies 3– and 4–year-old males began displaying significantly earlier than younger males.

The duration of the displaying period varied between males. Although considerable inter-individual variability was observed (range: 1–186 days of display), no difference was found between *macqueenii* and *undulata*.

## 4.3.3 Time of Laying

In 1992, the time of laying was recorded for 468 eggs (Saint Jalme *et al.* 1994). There was no difference between the two subspecies, so the data were combined. During the laying season sunset occurs between 18.00 and 19.00, and sunrise between 05.30 and 06.30. One per cent of eggs are laid in the morning after sunrise, 30 per cent between 15.00 and 18.00, and a large proportion (69 per cent) after sunset. For the last, it was not possible to record the precise time of laying. In many cases, when a clutch of two or more eggs is laid, the first egg is laid at the end of the afternoon and the second and third eggs most probably in the first half of the night.

## 4.4 Conclusions

Captive houbara bustards show a seasonal breeding pattern, with a laying period extending from January to July. Males begin displaying well in advance of any eggs laid, and continue to display long after the termination of laying. Our current belief is that an endogenous rhythm could have evolved in this species to regulate the annual reproductive cycle, comparable to that of species occurring in highly predictable environments. The primary proximate factor synchronizing sexual maturation could be the annual cycle of temperature. Rainfall could also act as a supplementary regulator, with abundant rainfall lengthening the reproductive period and allowing a second clutch to be laid. Some avian species inhabiting the arid interior of Australia are able to lay their first eggs within two weeks following significant rainfall (Dawson & Bartholomew 1968). Birds showing such a fast response to rainfall maintain gonadotropic activity in the hypothalamo-hypophysial axis under most conditions, and adults may reproduce even when moulting. For these species, rainfall may act as a proximate factor predicting when food resources will be favourable for the survival of the young. In our breeding system, food is given *ad libitum* and birds could start reproductive activities as soon as they have a reliable supply of protein, essential amino acids, vitamins, minerals and energy.

Delayed maturity is characteristic of houbara. Captive females are able to lay from 1 year of age, although the majority of females begin laying at 3. Males begin displaying at the age of 1, and the length of the display period increases with age. Reproductive parameters of females from the two sub-species are similar, although *macqueenii* generally display greater reproductive capacity; larger clutch size, and smaller inter-clutch intervals. The relatively limited number of birds does not permit us to determine if these differences are subspecific, or reflect only inter-individual variability. However, morphological differences between the two sub-species, and differences in patterns of display and in clutch size in the wild (Cramp & Simmons 1980), argue in favour of a real sub-specific difference.

While all birds used in these analyses were hand-reared, we still observed behavioural disparities between birds in the breeding unit; birds that fail to lay before 5 years of age are the most sensitive to stress. Generally, *undulata* appear to be more sensitive to disturbance and handling than *macqueenii*, and this could partly explain the difference observed between the sub-species in age of first breeding.

Clutch size is not related to age in both sub-species. Mendelssohn (1980) reports that females did not begin laying clutches of three eggs until 5 years old, whereas some of our *macqueenii* females laid four-egg clutches in their second year.

The analysis of laying dates shows two important phenomena:

the low degree of synchronization of sexual maturity in females, and late sexual maturation. If females in a population become sexually receptive in unison, there is little potential for individual males to monopolize multiple females. This is expecially true if each female is sexually active for only a brief period. With increasing degrees of asynchrony among members of one sex, the potential for individuals of the other sex to achieve multiple mating increases. Continuous long periods of sexual activity by males coupled with brief and asynchronous periods of receptivity by females will produce a strong skew in the Operational Sex Ratio (the average ratio of fertilizable females to sexually active males at any given time), and increase the potential for polygamy, suggesting a lek-type mating system (Emlen & Oring 1977).

Patterns of egg-laying in domestic bird species are governed by the interaction of two asynchronous factors; the cycle of follicular maturation and a cycle in the neuroendocrine mechanism which regulates the preovulatory release of luteinizing hormone (Sharp 1980). The neuroendocrine cycle is probably controlled by a circadian oscillator, and generates an 'open period' or 'gate' during which preovulatory release of LH may be initiated (Sharp 1980). According to this hypothesis, domesticated species such as the hen characteristically lay eggs during a 8–9-hour period of the day, in sequences, with the first egg being laid early in the day and successive eggs being laid later on successive days until the sequence is completed (Sharp 1980). In houbara, as 30 per cent of eggs were laid between 15.00 and 18.00, a symmetrical distribution will result in a ten-hour 'gate' between 14.00 and 24.00, with a laying peak between 18.00 and 20.00.

Time of laying and intervals between oviposition and between ovulation are very important parameters to be taken into account when artificial insemination is applied. Intervals between oviposition are more easily recorded than the interval between ovulation, and in many species both are very close (Sauveur 1988). In chickens, ovulation occurs about 24 hours before laying and fertilization occurs 15 minutes after ovulation (Sauveur 1982). Only a small quantity of spermatozoa is present in the upper part of the oviduct 15 minutes after insemination, and 24 hours are needed for 50 per cent of the spermatozoa to arrive in the area where fertilization of eggs occurs (Sauveur 1982). Therefore, inseminations must be performed at least two days before laying for successful fertilization of the chicken egg.

## Propagation of the Houbara Bustard

Within a clutch, houbara laid approximatively every 48 hours, but we do not know when ovulation occurred. Two lines of evidence suggest that ovulation occurs more than 48 hours before oviposition: eggs are laid two days apart within a clutch, with a shift of several hours between laying times, and no evidence has emerged to suggest that eggs can be fertilized by an insemination performed less than three days before laying (Chapter 5).

# 5

# Artificial Insemination and Natural Mating

P. GAUCHER, M. SAINT JALME and P. PAILLAT

## 5.1 Introduction

Information about the breeding behaviour of houbara in the wild is scarce and often contradictory. Some authors argue that houbara are monogamous with pair-bonds that last the duration of the breeding season (Ponomareva 1983), or that they are monogamous depending on food availability (Lavee 1988), whereas others describe them as polygamous (Collins 1983).

Published accounts indicate that houbara are difficult to breed in captivity, with only a small proportion of fertile eggs being laid, even though a large number of males may display (Mendelssohn *et al.* 1979; Mendelssohn *et al.* 1983; Ramadan 1989). Design and layout of the breeding cages have been suggested as determinants of the degree of successful fertilization. The most successful results have been achieved by permitting a limited number of individuals access to the largest possible area (Ramadan 1989).

As little specific information on natural breeding was available to guide us, we tried three types of management: (1) pairs; (2) small groups; and (3) groups with selective access to mates. The selective access techniques were derived from observations on the sub-species *fuertaventurae* of the Canary Islands (Collins 1983), which suggested that houbara are not monogamous, but polygamous with a dispersed lek. Males displayed at about the same location on consecutive days, and in relatively close proximity to each other. In common with normal lek behaviour, females could circulate among displaying males and choose to mate with any of them. Collins (1983) described the only observed copulation in this species, and concluded that since the female flew away immediately after copulation, no pair bond existed.

Poor results from our 'natural breeding' techniques prompted us to use artificial insemination. This is a very convenient tool when 'natural breeding' does not yield acceptable results. Variables of artificial insemination have been well studied in domesticated birds such as turkeys, guineafowl, chickens, mallard ducks and geese (Sauveur 1988), but relatively few studies have been published on most non-domesticated birds, except in the case of some falcons and raptors (Weaver & Cade 1991). Before efficient artificial insemination can be achieved, certain variables need to be defined: (1) the ability of the male to produce and give spermatozoa; (2) the sperm storage duration; (3) the minimum quantity of spermatozoa to be inseminated without reducing fertility; and (4) the timing of insemination in relation to the timing of laying. In chickens, inseminations performed less than five hours before or after oviposition have little chance of fertilizing eggs (Brillard 1982). Sperm storage duration is usually defined as the interval between the last insemination and the laying of the last fertile egg (Lake 1975), and is an important component of the female's fertile period, determining the frequency of insemination (Brillard 1982).

In this chapter we present results of 'natural breeding' and artificial insemination; including sperm production (volume of the semen, sperm concentration, and number of spermatozoa per ejaculate), and the capacity of the female to be fertilized (sperm storage duration, effect of quality of the semen on fertility, and hatchability levels).

## 5.2 Methods

### 5.2.1 *'Natural Breeding'*

We carried out 'natural breeding' trials on both sub-species. Pairs (n = 12) and small heterosexual groups (n = 8) were studied in *macqueenii*, and larger heterosexual groups with selective access to males (n = 7) in *undulata*. Pairs and groups of *macqueenii* were kept in separate communicating cages, the groups consisting of either one male with three females, or two males with two females. For *undulata*, a number of combinations of males and females, numbers of birds, and sizes of combined cages were tried. The basic design allowed between six and eight females to choose from between three and seven males, which were confined in individual communicating cages, and accessed by females

## Artificial Insemination and Natural Mating

through 'female doors' (Plate 5.1). Communicating doors were partially screened by a plywood board placed 50 cm. back from the opening, reducing visual contact between birds. The design was further modified to increase isolation of females, to enlarge the area available for males, and to allow the female to visit males, and then return to her own cage.

Mating cages were visited three times a day (07.00, 12.00, 18.00) to record male displaying activity; males were ranked in order of display persistence. All eggs were collected for artificial incubation.

### 5.2.2 Artificial Insemination

#### 5.2.2.1 Physiological Status of the Females

The body mass of females is a good indicator of physiological status, and is monitored weekly before laying. Since females start laying when body mass is at a maximum, laying body mass of females in previous years indicates when we should begin to inseminate. For females laying for the first time, we begin to inseminate when the body mass has increased by 15 per cent above the minimum, which occurs between September and November.

#### 5.2.2.2 Semen Collection

The method used for semen collection involves voluntary participation by the male, and is only possible with hand-reared birds. A dummy female mounted in a sitting position is presented to a displaying male (Plate 5.2). He approaches, performing the pre-copulation display, and pecks the dummy female's head before mounting its back (Fig. 5.1). He then sits back on his tarsi, and begins a trampling motion. At the same time the wings are spread with the tips touching the ground to keep balance. The handler should lift the glass collection dish to touch the cloacal region: this triggers ejaculation, which is accompanied by a quick down-flap of the tail. Semen can also be collected directly from the cloaca (Plates 5.3 & 5.4). The first method is used with highly sensitive birds which object to handling, but semen may be lost in the feathers surrounding the cloaca when the male ejaculates. To collect with the second method it is necessary gently to massage both sides of the back of the male when he is mounting the female in order to encourage him to ejaculate with the tail held high, making the cloaca easily accessible. Only one person is needed to collect semen using this method, and so stress experienced by the birds is reduced.

The semen is immediately transferred into a 1 ml. vial which contains 0.2 ml. of commercial diluent (I.M.V., Laigle, France; guineafowl diluent). It is then stored at ambient temperature and in the dark.

### 5.2.2.3 Sperm Analysis

Immediately after collection, semen is brought to the laboratory and analysed for concentration and motility of spermatozoa. Between 1989 and 1991 we assessed sperm quality qualitatively using a microscope. When a sample of semen was considered good in relation to average concentrations and motility of spermatozoa, it was used for insemination.

In 1992 a quantitative study was undertaken (Saint Jalme *et al.* 1994). We collected semen in 0.2 ml. of diluent. Once in the laboratory, 0.02 ml. of the solution ejaculate and diluent was diluted ten-fold for analysis of sperm concentration, number of spermatozoa per ejaculate, and motility of spermatozoa. For these analyses a Thomas' cell was used.

Motility of spermatozoa was scored on the following scale: (1) no motility; (2) less than 50 per cent of spermatozoa showing activity; (3) more than 50 per cent of the spermatozoa mobile but majority of individual tracks only circular and local; and (4) more than 80 per cent of the spermatozoa mobile and individual movement spread over several counting squares.

### 5.2.2.4 Insemination

We perform inseminations between 07.00 and 10.00, and inseminate within 30 minutes after collection (Plate 5.5). The cloaca is opened with special forceps. Using a head-torch, the entrance to the oviduct can then be seen on the left side of the cloaca. If the cloaca is full of faeces, physiological serum is used to evacuate the faeces, and expose the entrance to the oviduct. The size of the entrance is related to the proximity of oviposition. In a female ready to lay, it is possible to see the entrance of the oviduct by pinching between the thumb and index finger the superior lips of the cloaca, and everting them slightly. This method is prefer-

## Artificial Insemination and Natural Mating

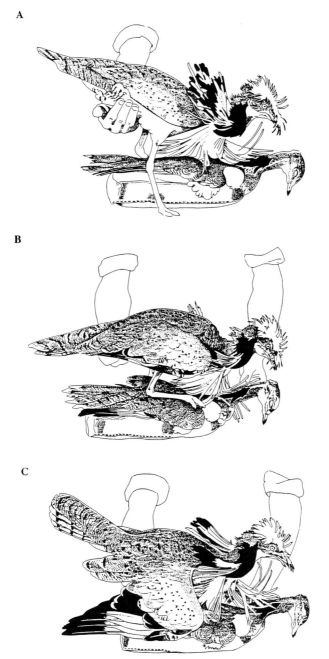

Figure 5.1 Male mounting a dummy female during the semen collection procedure

able as it is less stressful to the female than the technique requiring forceps. A plastic tube is inserted about 1 cm. inside the entrance to the oviduct, and the semen is introduced by slowly applying gentle pressure on the piston of the syringe. One person should hold the female and one person inseminate. We carry out all inseminations in the females' cages.

Between 1989 and 1992 we performed inseminations at intervals of three to 30 days before laying, to investigate sperm storage duration inside the female reproductive tract (Saint Jalme et al. 1994). In 1992, the intervals between the two inseminations were between four and 15 days, so that females received between one and three inseminations during the 15 days before laying. The quantity of spermatozoa inseminated ranged for each insemination between one and 50 million spermatozoa.

As an extension of the 1992 study, in 1993 females were inseminated every four days, and when possible with no less than 10 million spermatozoa in each insemination.

## 5.3 Results

### 5.3.1 'Natural Breeding'

Although sample sizes are too small to draw any definite conclusions, the largest proportion of laying females (89 per cent) was from groups consisting of two or three females together with one or two males, housed in a minimum of ten communicating cages. The proportion of females laying fertile eggs was also highest in this group. However, some of the females chosen for these trials were selected on the basis of their good performance in the preceding year.

Within the paired group, fewer females (58 per cent) laid, 70 per cent of them produced fertile eggs, and 21 per cent of the eggs were fertile. Within the heterosexual group with selective access to males, almost all males displayed (94 per cent, n = 32), although males showed considerable inter-individual variation in the mean number of days on which they displayed (mean = 79.9 days of display per male within each group, SD = 47.5, n = 7 groups).

The proportion of females laying eggs in these groups (56 per cent) was the same as in the paired treatment, although fewer females laid fertile eggs (52 per cent). In general, the maximum

*Artificial Insemination and Natural Mating*

**Table 5.1 Summary of 'Natural Breeding' Results (1990–1993)**

|  | N | No. laying females | No. fem. laying fert. eggs | No. fert. eggs |
|---|---|---|---|---|
| Pairs | 12 | 7 (58%) | 5 (71%) | 10/47 (21%) |
| range |  | (-) | (-) | (0 – 43) |
| Group$^1$ | 18 | 16 (89%) | 12 (75%) | 67/149 (45%) |
| range |  | (80 – 100) | (80 – 100) | (20 – 67) |
| Group$^2$ | 48 | 27 (56%) | 14 (52%) | 45/149 (30%) |
| range |  | (17 – 83) | (0 – 100) | (0 – 56) |

Group$^1$ = *C.u. macqueenii*; small groups.
Group$^2$ = *C.u. macqueenii* & *undulata*; larger groups.

number of fertile eggs (56 per cent) was obtained from groups showing highest display rates.

Before breeding activity commenced, most females remained in one flock, or in smaller groups in a single cage. If the 'female door' is in position, allowing access to the males, females may congregate in the cage of a non-displaying male. Upon reaching laying condition, females seek isolation and become aggressive, particularly towards other females. The use of screens as a visual barrier is important, as without them almost half the clutches were abandoned. Laying females usually select cages furthest away; males' cages are only selected if the male does not display. If females lay in males' cages, male behaviour seems to be inhibited by the presence of the female.

For a summary of 'natural breeding' results (1990 to 1993), see Table 5.1.

### 5.3.1.4 Analysis of Paternity in *undulata*

In experiments involving several males it is difficult to identify the parents of each egg. We used DNA fingerprinting to determine the paternity and maternity of hatchlings from fertile eggs. In four eggs collected from two groups of two females in separate cage blocks, paternity was from the same male in each block. These males were also the most persistent displayers of their block. In another cage block, three females were found to have mated with the oldest best displayer. In a fourth cage block, three males fathered eggs from four females: three females were fertilized by one male, while the fourth copulated with two males. Two of the three males copulated with two females. These males displayed most among the five males in the cage block.

## Table 5.2 Results of Artificial Insemination, 1989–1993

### C. u. macqueenii

|      | No. of females | No. of eggs | No. fertile eggs | Fertility level (%) |
|------|----------------|-------------|------------------|---------------------|
| 1989 | 12             | 34          | 19               | 55.9                |
| 1990 | 19             | 98          | 51               | 52.0                |
| 1991 | 29             | 143         | 77               | 53.8                |
| 1992 | 28             | 228         | 150              | 65.8                |
| 1993 | 35             | 282         | 231              | 81.9                |

### C. u. undulata

|      | No. of females | No. of eggs | No. fertile eggs | Fertility level (%) |
|------|----------------|-------------|------------------|---------------------|
| 1989 | 7              | 19          | 9                | 47.4                |
| 1990 | 16             | 62          | 40               | 64.5                |
| 1991 | 4              | 15          | 11               | 73.3                |
| 1992 | 24             | 143         | 107              | 74.8                |
| 1993 | 36             | 249         | 222              | 89.2                |

No. of females: Number of laying females inseminated.
No. of eggs: Number of eggs laid when at least one insemination was performed between 3 and 15 days before laying.

## 5.3.2 Artificial Insemination

### 5.3.2.1 Sperm Production

Results obtained in 1992 (Saint Jalme *et al.* 1994) indicated a mean ejaculate volume of between 0.07 and 0.08 ml. (range: 0.01 – 0.3 ml.). The mean concentration of spermatozoa was about $350 \times 10^6$ spermatozoa ml$^{-1}$ in each ejaculate (range: $0.3 \times 10^6$ – $4515 \times 10^6$). For all sperm production variables, there is substantial inter- and intra-individual variability; no difference was found between sub-species. The average number of spermatozoa per ejaculate was about $20 \times 10^6$. As for sperm concentration, large intra- and inter-individual variation was found, with values ranging from 10,000 to $405 \times 10^6$ spermatozoa per ejaculate. The number of spermatozoa per ejaculate collected per week is highest when sperm is collected at least every three days, but decreases when the collection frequency increases.

### 5.3.2.2 Fertility results

The mean fertility level from 1989 to 1991 was 62.4 per cent in *undulata* and 53.5 per cent in *macqueenii*; this difference is not

## Artificial Insemination and Natural Mating

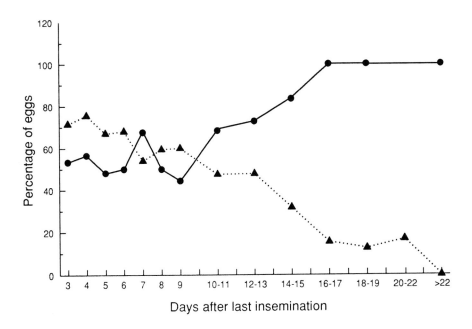

**Figure 5.2** Effect of sperm storage *in vivo* on fertility (triangles) and embryo mortality (circles) in houbara; n = 706 eggs

significant. Combined fertility levels increased in 1992 (69.3 per cent) and again in 1993 (85.3 per cent) as results of the 1992 study were applied (Table 5.2). Substantial inter-individual variability was observed in every year: for example, in 1993, 12 per cent of the females showed less than 25 per cent fertility, 8 per cent between 25 and 50 per cent, 22 per cent between 50 and 75 per cent, and 58 per cent more than 75 per cent.

### 5.3.2.3 Estimation of Sperm Storage Duration Within the Oviduct

Experiments showed that eggs cannot be fertilized if the insemination is performed less than three days before laying (Fig 5.2). Highest fertility rates were obtained when birds were inseminated about four days (range: three – six) before laying (76 per cent). If inseminated between seven and nine days before laying, the fertility level was lower (60 per cent). A delay of 10 days between insemination and laying meant fertility levels decreased to less than 50 per cent. No difference in sperm storage duration was

**Table 5.3 Relationship Between Number of Spermatozoa Inseminated During Three Time Intervals Before Laying, and Fertility Level of Eggs as a Percentage of Total Eggs (n = number of eggs)**

| No. sperm (millions) | Time interval | | | | | |
|---|---|---|---|---|---|---|
| | 3–15 days | | 3–10 days | | 3–6 days | |
| | n | % | n | % | n | % |
| 0–5 | 25 | 28.0 | 71 | 45.1 | 148 | 50.7 |
| 5–10 | 45 | 46.7 | 71 | 56.3 | 67 | 68.7 |
| 10–15 | 56 | 51.8 | 51 | 68.0 | 46 | 82.6 |
| 15–20 | 33 | 60.6 | 32 | 68.8 | 34 | 82.4 |
| >20 | 176 | 81.8 | 110 | 82.7 | 40 | 85.0 |

found between sub-species. Median sperm storage duration was 10 days and maximum sperm storage duration 22 days.

### 5.3.2.4 Effect of Number of Spermatozoa Inseminated on Fertility

The 335 eggs obtained in 1992 were classified according to the number of spermatozoa inseminated, with a cluster interval of 5 million spermatozoa (Table 5.3). The number of spermatozoa inseminated was summed for each of the three following time intervals before laying: 3–6 days, 3–10 days and 3–15 days. Percentage of fertile eggs was calculated from all the eggs, for each of the three time-periods. The best fertility results were obtained when more than 10 million spermatozoa were inseminated within the period 3–6 days before laying.

## 5.4 Conclusions

Birds in the 'natural breeding' trials bred in both monogamous and promiscuous situations, suggesting that mating strategies are flexible. Males displaying most persistently are more likely to fertilize eggs. Females in laying condition appear to be intolerant of the presence of other females, and require a large amount of space for successful reproduction. Management of houbara in cages should also allow females to have continuous access to a displaying pool of males.

Natural copulations in domestic species such as chickens and guineafowl usually produce a high level of fertility (more than

90 per cent), and the main aim of artificial insemination is to reduce the number of males needed to fertilize females without compromising fertility levels. In houbara, the low degree of fertility in our 'natural breeding' trials (0 to 67 per cent fertile eggs) is probably due in part to inadequate management techniques, resulting from our poor understanding of the breeding behaviour of this species. Cages could be too small, or there may be a lack of sexual synchronization between males and females owing to considerable variability in length and timing of the reproductive season. In the wild, females may have access to a larger number of potential mates, allowing them to choose one that is sexually mature at the moment they are ready to lay, whereas in the captive breeding unit females have access to a relatively small number of males. Artificially inseminating houbara increased the mean fertility level to 85 per cent in 1993, a mean increase of 50 per cent above that achieved in 'natural mating' trials.

Results from experiments in 1990 and 1991 indicated that we should increase the insemination frequency in 1992. Precise control of semen quality allowed us to select males most likely to produce the best results. No difference in sperm production was found between the sub-species. The mean volume of ejaculate obtained in houbara is very close to that found in guineafowl of about the same body mass (Sauveur 1988: 0.05 ml.-0.25 ml.), however, sperm concentrations recorded from houbara are much lower than in guineafowl (Sauveur 1988: $5 \times 10^9$-$8 \times 10^9$ sp/ml.). Intra-individual variation in sperm concentration and in number of spermatozoa per ejaculate was largely due to seasonal variation, and in the frequency of semen collection. We obtained the best number of spermatozoa per ejaculate by collecting every three days. Similar results have been obtained with chickens (de Revier 1982; Sauveur 1988). Therefore, the mean quantity of spermatozoa produced per week in fully sexually developed houbara is about $165 \times 10^6$, which is very low in comparison with guineafowl or chickens, which produce respectively $4 \times 10^9$ (de Revier 1982) and more than $20 \times 10^9$ spermatozoa per week (Sauveur 1988).

Results of sperm storage duration '*in vivo*' were obtained by summarizing all data without taking into account the quality of inseminations. This produces an underestimate of the median sperm storage duration. We found a positive correlation between fertility level and quantity of spermatozoa inseminated, and a direct relationship between sperm storage duration and number of sperm inseminated. Best fertility results were obtained in 1993 by inseminating more than 10 million spermatozoa, between three and six days before laying.

When houbara males are fully developed, best fertility and hatchability can be achieved by collecting semen every three days (mean of 39 million spermatozoa/ejaculate), and by inseminating 10 million spermatozoa (an average of 4 females/ejaculate) every four days; all eggs should receive at least one insemination between three and six days before oviposition. Application of these methods in our flock resulted in a fertility level of 85 per cent.

# 6
# Incubation

P. GAUCHER, P. PAILLAT and M. SAINT JALME

## 6.1 Introduction

According to Povlawski (1965) the incubation period in houbara is 28 days. Mendelssohn *et al.* (1983) found eggs hatched after 23 or 24 days when they were artificially incubated at 37°C. Because incubation parameters (temperature, humidity, mass-loss, etc.) of houbara eggs were unknown, we collected data on natural conditions of incubation in order to improve artificial incubation techniques.

The basic principle of artificial incubation is to provide a stable temperature, adjust humidity as circumstances dictate, and to follow a designated egg mass-loss curve for the species. Such curves have been extensively worked out for falcons and other raptors (Weaver & Cade 1991). One of the most important factors in ensuring hatching success is the ability to control mass loss in line with the designated curve. Mass-loss curves for most captive species have been well documented. An egg loses mass by evaporation during the incubation period. This water loss is related to temperature, relative humidity and flow of air around the egg. The rate of loss is also determined by shell porosity. It is generally recognized that an egg must lose a minimum of 11 per cent of its initial mass to have any chance of hatching, and 14 per cent is usually ideal. Because nothing was known about these parameters in houbara, it was necessary to establish the true natural curve and base our control of artificial incubation parameters on this curve.

Our choice of incubators was based on past experience, and the need to provide a degree of adaptability depending on the results achieved. Several types of incubators were tested: 'forced-air' and 'still-air' incubators with a capacity of holding about 30 eggs were initially selected. These small-capacity incubators allow considerable control over individual eggs, which is desirable as there is great variability in eggs from different females.

Incubation procedures are also dictated to some degree by the management of the breeding flock, and behavioural patterns of the laying females. Because our policy is to maximize the breeding potential of our houbara, we followed a procedure of 'egg-pulling', which involves removing the fresh egg as quickly as possible. This minimizes the possibility of the female going 'broody', and in this way we can increase the number of eggs laid within the breeding season.

## 6.2 Methods

### 6.2.1 *Natural Incubation*

Natural incubation parameters were studied in 1990 using dummy eggs placed in a fixed position within the nest cup. These eggs had two thermistors implanted at the point of contact with the brood patch, and temperature was recorded and stored on a logger for analysis (Fig. 6.1). Video recording of the incubating female was also carried out throughout the incubation period (Schulz *et al.* 1991). As well as this, eggs incubated by females were removed from the nest every two days and weighed, so that a mass-loss curve could be plotted. Between 1992 and 1993, 30 females incubated 65 eggs. Hatchability of these eggs was compared to those incubated artificially.

### 6.2.2 *Artificial Incubation Procedures*

After testing several incubators, we chose VOM1 (Schumacher, Germany); an incubator with a capacity of 30 eggs. Stability and control of temperature in the incubation room is achieved using a pre-conditioning room (described in Chapter 3). Temperature is maintained at 21.5°C, and humidity is around 30 per cent. The number of air changes and degree of positive pressure can be altered by adjusting the efficiency of fans and controlling the outlets (Plate 6.1).

We collect eggs twice daily between 07.00 and 19.00. They are lifted using a surgical glove, and placed in a container lined with foam plastic. The egg is numbered and taken to the egg-cleaning room. Measurements of length, breadth and mass are entered on to an individual data sheet containing a percentage mass-loss graph, and columns for recording incubator changes,

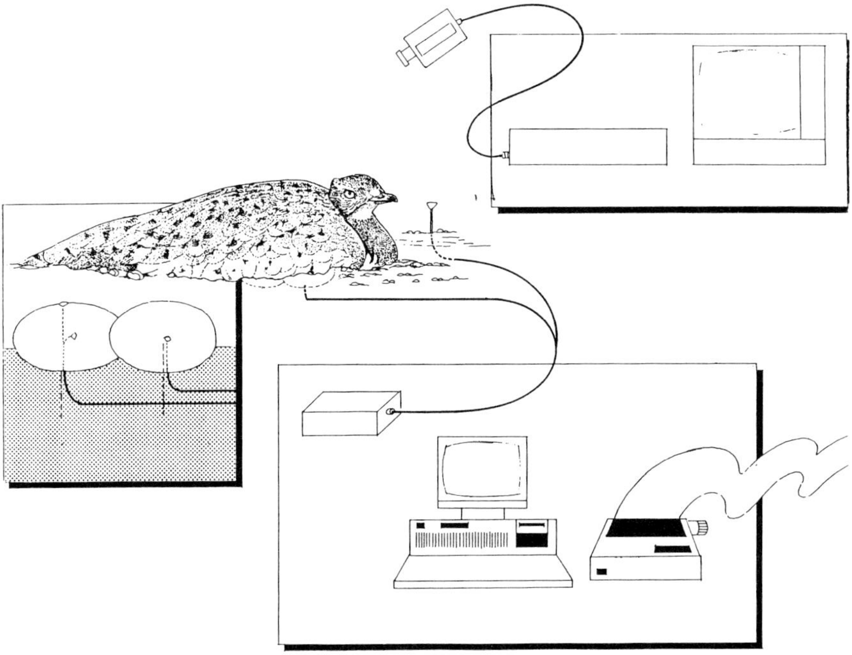

## Incubation

Figure 6.1 Apparatus for recording incubation parameters

mass loss and comments (Fig. 6.2). Eggs are dipped into an egg sanitant (ANTEC, chickguard), and are then placed on a plastic egg tray to dry.

A cold egg should not be brought up to incubating temperature too suddenly. If the eggs are cold when collected, we let them adjust to room temperature over a period of two hours, at 22.5°C. They are then weighed and placed in a pre-incubator, which is set to operate at 30°C; the eggs are brought up to this temperature over a two-hour period. This procedure also avoids potential adverse affects caused by introducing a cold egg into a warm incubator, thus changing the conditions within the incubator. After the four-hour adjustment period the eggs are placed into a 'dry' incubator, operating at 37.6°C, and with a humidity of less than 30 per cent. If the eggs are collected 'hot', they can be placed directly into the incubators, after the disinfection procedure.

Humidity is controlled by placing none, one, or two glass dishes of water on the floor of the incubators; these provide humidities of about <30 per cent ('dry'), 45 per cent and 60 per cent respectively. Eggs are transferred between incubators set at different humidities, to achieve a rate of mass loss that tracks

## Propagation of the Houbara Bustard

FEMALE    Date laid: 18/03/92                              EGG
Brown 12   Weight: 53.68   Length: 56.5   Breadth: 42.3   No. 188

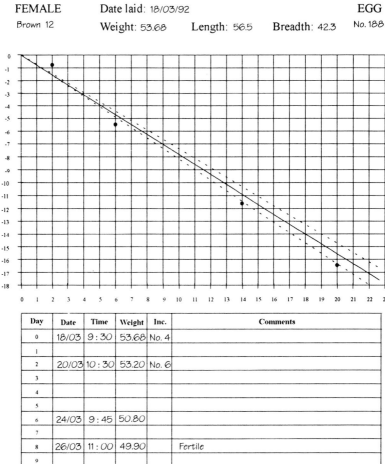

| Day | Date  | Time  | Weight | Inc.  | Comments              |
|-----|-------|-------|--------|-------|-----------------------|
| 0   | 18/03 | 9:30  | 53.68  | No. 4 |                       |
| 1   |       |       |        |       |                       |
| 2   | 20/03 | 10:30 | 53.20  | No. 6 |                       |
| 3   |       |       |        |       |                       |
| 4   |       |       |        |       |                       |
| 5   |       |       |        |       |                       |
| 6   | 24/03 | 9:45  | 50.80  |       |                       |
| 7   |       |       |        |       |                       |
| 8   | 26/03 | 11:00 | 49.90  |       | Fertile               |
| 9   |       |       |        |       |                       |
| 10  |       |       |        |       |                       |
| 11  |       |       |        |       |                       |
| 12  |       |       |        |       |                       |
| 13  |       |       |        |       |                       |
| 14  | 01/04 | 9:05  | 47.41  |       |                       |
| 15  |       |       |        |       |                       |
| 16  |       |       |        |       |                       |
| 17  |       |       |        |       |                       |
| 18  |       |       |        |       |                       |
| 19  |       |       |        |       |                       |
| 20  | 07/04 | 10:10 | 44.88  |       | Pipped                |
| 21  |       |       |        |       |                       |
| 22  | 09/04 | 6:45  |        |       | Hatched  Ring: Yellow 8 |
| 23  |       |       |        |       |                       |

**Figure 6.2 Example of a mass-loss recording sheet**

that plotted for naturally incubated eggs. Egg-turning is carried out automatically every hour.

Up till the nineteenth day of incubation, eggs are cooled daily at room temperature inside the incubating room (21.5°C) for 20 minutes. This is when candling and weighing is done, and transfer between incubators set at different humidities occurs. After 19 days we stop turning the eggs, and transfer them into a 'dry' incubator set at 37.6°C, and with additional oxygen at 22 per cent; atmospheric oxygen is lower at the altitude of the NWRC (1450 m.), and this may cause embryos to be weaker and less likely to hatch successfully. Pipping eggs are transferred to the hatcher, which is maintained at 70 per cent humidity and 37°C temperature. After hatching, the navels of chicks are cleaned with an antiseptic cream containing vitamin A, and chicks are placed in a drier until they are transferred to the rearing unit.

Infertile and dead eggs are identified by candling twice a week, and are removed from the incubators. All these eggs are opened in order to confirm their status (fertile or infertile), and to evaluate the age at which they had died.

Because we wished to achieve greater hatching synchrony, we modified incubation techniques in 1993: between 8 and 28 March, 94 eggs were stored in the egg-storage room at 8°C, and 70 per cent humidity. They were turned twice a day, and transferred into incubators twice a week. During the remainder of the year, eggs were transferred into incubators immediately after collection.

## 6.3 Results

### 6.3.1 *Natural Incubation*

Naturally incubated eggs pipped at 21 days and hatched at 22 days. The average mass loss during incubation from seven fertile eggs which hatched successfully was 17.4 per cent (range: 15.4 to 20 per cent). The first egg is not constantly attended by the female, but is incubated only during the hottest period of the day (between 11.00 to 15.00) and the coldest period of the night (between 01.00 to 06.00), presumably to protect it from extremes in temperature. Attendance at the nest increases from the laying of the second and third eggs over the next 10 days. The female leaves the nest for approximately one hour twice a day, between 05.00 and 09.00 in the morning and 16.00 to 19.00 in the evening, for feeding, preening and other activities.

Females turned eggs, on average, 13 times a day. For technical reasons it was not possible to record this information during the night. Incubating females collected small stones (4–5 mm.) that were accessible from her prone position, placing them on her back or around her sides, and eventually creating a ring of stones around the clutch. The reason for this behaviour is unknown.

After hatching, the female removes the shell fragments from the nest site as far as is possible within the cage area. She remains on the nest for about 24 hours after hatching before moving off with her chicks. Eighty per cent of naturally incubated eggs hatched (Table 6.1).

### 6.3.2 Artificial Incubation

The majority of artificially incubated eggs hatched after 22 days and some after 23 days. No significant differences were found in levels of fertility or in hatchability between the two sub-species. The mean hatchability level obtained in 1993 was 65.5 per cent. Deaths of embryos during the two first days of incubation cannot be attributed to incubation techniques (Sauveur 1988). If early embryonic failure is added, the hatchability level decreases to 59 per cent (Table 6.1).

A relationship was found between the ability to hatch and the interval between insemination and laying: no differences were found when inseminations occurred less than 10 days before laying, but when inseminations were performed more than 10 days before laying, we observed an increase in embryo mortality (Fig 5.2).

### 6.3.3 Embryo Mortality

Embryo mortality of artificially incubated eggs was concentrated at the beginning and end of incubation (Table 6.1): from 170 embryos that died during incubation, 44 per cent died in the first

**Table 6.1 Embryo Mortality: Comparison of Natural and Artificial Incubation**

|  | Age of death (days) | | | | | **Total** |
|---|---|---|---|---|---|---|
|  | 1–2 | 3–5 | 6–17 | 18–22 | Hatched | |
| Natural | 4 (6%) | 4 (6%) | 3 (5%) | 3 (5%) | 51 (78%) | 65 |
| Artificial | 41 (10%) | 33 (8%) | 21 (5%) | 75 (17%) | 245 (59%) | 415 |

*Incubation*

**Table 6.2 Comparison Between Levels of Embryo Mortality Associated with Two Incubation Techniques: Eggs Incubated Immediately After Laying (A); Eggs Transferred to Incubator Between Two and Four Days After Laying (B)**

|   | Age of death in days | | | | | |
|---|---|---|---|---|---|---|
|   | 1–2 | 3–5 | 6–17 | 18–22 | Hatched | **Total** |
| A | 25 | 13 | 11 | 44 | 169 | 262 |
|   | 10% | 5% | 4% | 17% | 65% | |
| B | 12 | 17 | 5 | 17 | 43 | 94 |
|   | 13% | 18% | 5% | 18% | 46% | |

four days and 44 per cent in the last four days. The remaining 22 per cent died between six and 17 days of incubation.

The level of hatchability obtained from eggs incubated under the females was 19 per cent better than from artificially incubated eggs. This difference is largely due to higher levels of mortality in the four days before hatching in artificially incubated eggs (5 per cent in naturally incubated eggs and 17 per cent in artificially incubated eggs).

The modification of incubation techniques in 1993 indicated that storing eggs before transferring them to incubators resulted in 19 per cent lower hatchability (Table 6.2), with higher mortality between three and five days of development (unstored: 5 per cent, stored: 18 per cent).

## 6.4 Conclusions

Eggs incubated by females pipped at 21 days and hatched at 22 days. The longer incubation period observed by Mendelssohn *et al.* (1983) is probably due to a lower incubation temperature. The mean mass loss during incubation is about 17 per cent. Hatchability of artificially incubated eggs was 19 per cent lower than for naturally incubated eggs, largely due to higher levels of mortality in the four days before hatching. Hatchability decreased if inseminations had been performed more that 10 days before laying. A decrease in viability of eggs fertilized by spermatozoa that have been stored for a long time has been shown in numerous species, such as turkeys, chickens, mallards, and Japanese quail (Birkhead & Moller 1992). In chickens, the maximum storage duration is 35 days, but the probability of an egg hatching after 20 days of sperm storage is low (Lodge *et al.* 1971). Nalbandov and Card (1943) have shown that the older the sperm the earlier the stage of development at which mortality occurs. The mean hatchability level obtained for houbara in 1992 (49.2 per cent, or 55.7 per cent excluding early dead embryos) was relatively low compared to 1993. Results showed that by decreasing the interval between insemination and egg laying, and increasing the quantity of spermatozoa inseminated, we can expect as much as 70 per cent hatchability (excluding embryos that die before three days of development). When eggs were transferred into an incubator as quickly as possible after laying, we obtained a hatching rate of 70 per cent.

The high mortality between one and two days is difficult to interpret. In the domestic chicken, the passage of the yolk from the oviduct takes 24 hours, and development of the embryo commences during this stage. After laying, further development will be arrested by the cooling of the egg. However, some breeds of chicken take over 24 hours to lay an egg and their fertility is very low. Early mortality is not thought to be directly related to problems in artificial incubation. The most susceptible stages of development are the formation of the primitive line (10–16 hours) and the positioning of the vitelline veins during the second day of life. The first stage is very sensitive to egg-storage conditions. While parameters affecting the second stage are not well known, insufficient warmth and too long a storage period may often be the cause of death (Sauveur 1988).

The high number of deaths at the end of incubation is also difficult to interpret. Incubation at relatively high altitude (Wisschedijk & Rahn 1981, in Sauveur 1988), with a low concentration of ambient $O_2$ and $CO_2$, could be partly responsible for these deaths.

While our incubation methods ensure a sizable production of chicks, there is still room for improvement of both equipment and procedures. In 1993, results from captive breeding improved markedly. A total of 670 eggs were laid by 75 females. The mean fertility level achieved through artificial insemination was 85.3 per cent, an increase of 15.8 per cent compared to 1992. The hatching rate (65.5 per cent including early deaths) increased by 12 per cent from values recorded in 1992, although we still have a relatively high level of embryo mortality at the end of incubation (18 per cent).

# 7

# Rearing

P. PAILLAT and Y. VAN HEEZIK

Procedures used in chick rearing will influence the suitability of chicks either for breeding stock, or for release into the wild. Important factors include diet, which influences physical development, and the social environment during rearing, which may determine future survival skills and social integration once in the wild. Two methods were used to rear chicks: (1) by females caged in the breeding unit, and (2) by hand.

## 7.1 Rearing by Females in the Breeding Unit

The female covers the chicks for the first 24 hours after hatching, and no supplementary food is given. Subsequently, live mealworms are sprinkled on the ground in her vicinity at two-hourly intervals during the day. These are presented to the chicks by the female, which utters low calls that stimulate a feeding response. Food presented in this way is coated in saliva by the female.

On the second day, young crickets (5–10 mm.) are added to the mealworms. These are chilled in a freezer beforehand to prevent them from escaping. After two or three days chicks become very adept at receiving food from the female's bill, and she encourages them to collect their food from the ground by dropping the mealworms or crickets before they have a chance to grasp them from her bill (Plate 7.1). After five or six days chicks begin to feed independently, although some food is still received from the female.

At two months of age the proportion of insects in the diet is reduced, and chicks are slowly shifted to a rearing ration of pellets, as well as alfalfa from the cultivated plots within their cage. Chicks raised by the female are very sensitive to human disturbance, even though the females are quite tame. Consequently, these chicks are not suitable for future captive breeding.

## 7.2 Hand-rearing

After hatching, chicks are moved to a dryer (37°C) for three to four hours, antiseptic cream containing Vitamin A is applied to their navels, and they are ringed. Once dry they are transferred to the rearing unit and placed in a brooding box made from Plexiglas (78 cm. × 50 cm.). This has no lid or base, and is placed on a table covered with sterilized sand, with a layer of fine plastic mesh on top (Plate 7.2). The mesh is to prevent chicks from ingesting excessive amounts of grit, which may result in an impaction of the digestive tract. Plexiglas is a smooth material that can be cleaned easily; boxes are cleaned every few weeks with a solution of ammonium IV. A space is left between the rearing boxes to allow them to be moved to a new position when the sand becomes contaminated by stale food or faeces. The sand is cleaned every two weeks and sprayed with a solution of ammonium IV. The main aim of the sanitary procedures is to maintain a balance of micro-organisms within the rearing unit that the chicks become immune to, but to avoid introducing high levels of contaminants. To this end, it is most important to control which organisms enter the rearing unit.

Infra-red heaters are suspended from a rail running around the wall 150 cm. above the rearing table and 35 cm. from the wall; this rail contains electrical sockets placed every 150 cm., from which heating lamps are suspended. This system allows the lamp to be relocated to any position along the rail. Temperature is controlled by raising or lowering the lamp. Chicks should have access to an area warmed to 37°C. The atmosphere should be warm, but with adequate ventilation, though preferably not directly around the chicks.

When hand-rearing, four or five chicks of the same age are placed in each box, since interactions between chicks stimulate a feeding response. Approximately 24 hours after hatching, mealworms and alfalfa are offered to the chicks. Meals are offered every half hour. Using blunt forceps, dead mealworms that have been dunked in a highly absorbable solution of calcium (Collocal D.P.) and vitamins are placed in front of each chick's bill. After one day young crickets and alfalfa are added to the mealworms, and grit must be provided to allow digestion of the chitin (Plate 7.3). The behaviour of the female is imitated by dropping the food in front of the chicks, to encourage them to feed by themselves. At this time chicks are started on pellets (Special Dietary Services) that have been softened and broken up in a solution of calcium and vitamins. Even after the chicks are feeding independently we continue to give some food by hand to maintain close human contact (Plate 7.4).

Between 2 to 3 weeks of age chicks are transferred into small rooms that have a sandy floor and a lamp in one corner for warmth. Access to an exterior pen is made available when chicks are old enough to cope with large variations in temperature (at 1 to $1\frac{1}{2}$ months). Windows in the rooms permit sun to shine through in patches, as sun is required for vitamin D metabolism. The outside pen is lined with tensioned shade-cloth. During the day older chicks are free to move both in and outside.

When transferred, chicks are usually feeding independently on a mixture of pellets, alfalfa and mealworms, although some may need some initial encouragement to begin feeding. Food and water is provided *ad libitum*, once a day, and dishes are removed at night for cleaning. By the end of the third month the diet consists only of dry pellets. Chicks are very curious and will peck at and swallow any unusual items they find in their cages; therefore care should be taken to keep these areas free of any potentially dangerous objects. Ideally less than 10 chicks are housed per room, and they are kept here until 2 to 3 months of age. After transfer of chicks to either the breeding unit, or for translocation, the rooms are cleaned and disinfected before the next group arrives.

Older chicks in flocks are nervous, so that handling should be minimized to avoid deaths from trauma. If chicks show signs of illness, it is necessary to judge whether attempting to catch and treat the chicks is absolutely vital, as any disturbance of the flock is associated with a high likelihood of an accident or death.

## 7.3 Growth of Hand-reared and Naturally Raised Chicks

Although the majority of young houbara at the NWRC show apparently normal growth (Fig 7.1), two problems have been identified in our chicks: (1) 'slipped wing' syndrome, and (2)

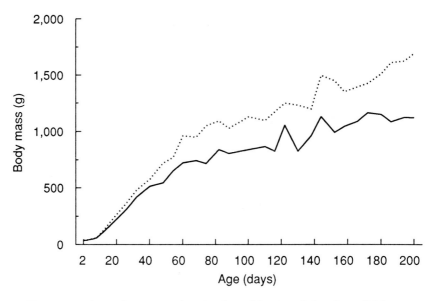

**Figure 7.1** Growth curves of male (dotted line) and female (solid line) *macqueenii* chicks

bone deformities occurring between the age of 7 and 20 days. 'Slipped wing syndrome' could be linked with an over-proteic or over-caloric diet. It is a problem frequently encountered with chicks, initially occurring at about 15 days of age when the primaries adopt malposition by turning out from the body. Both wings may be affected. In other species, it has often been assumed to be a nutritional problem which could be caused by too much protein in the diet, which causes an unnatural growth pattern. Chicks which display slipped wings have them taped into a natural position: the correct position is usually assumed within a few days, with no recurrence of the symptoms. Problems concerning bone deformities are discussed in Chapter 8.

In 1993, the calcium intake of houbara chicks was raised to 5 per cent of the chicks' diet; a value much higher than for any other species, where requirements range from 1.5 to 2.5 per cent. Calcium must be administered in a highly absorbable form (Collocal D.P.) to be effective. Increased doses of vitamins and trace elements were also provided. The following vitamin complex was added to the drinking water: Vitamin A, 15000 i.u.; Vitamin $D_3$, 1000 i.u.; Vitamin E, 10 mg.; Vitamin $B_1$, 10 mg.; Vitamin $B_2$, 5 mg.; Vitamin $B_6$, 3 mg.; Nicotinamide, 35 mg.; Panthenol, 25 mg.; Vitamin $B_{12}$, 50 mcg.; Chlorocresol, 1 mg. This

*Rearing*

treatment succeeded in almost totally eliminating bone deformities.

## 7.4 Conclusions

The need for specific research on growth and development of chicks as a function of their diet is apparent. Further research on houbara chick feeding should define the optimum diet, compromising growth rate in order to prevent 'slipped wing' syndrome, but without stunted growth. In addition, as production of chicks for release into the wild improves, research on chick-rearing methods designed to optimize their survival potential after release will be undertaken.

# 8
# Pathology and Veterinary Care

S. OSTROWSKI, A. GRETH and I. MIKAELIAN

## 8.1 Introduction

Diseases and pathology of the houbara bustard were poorly documented. This lack of knowledge posed a serious threat to the success of the captive-breeding project at the NWRC. Many pathological events occurred during the first years of captivity of this species, allowing the collection of data on virology, bacteriology, histology, parasitology, immunology and therapeutics. To follow the pathology, every sick bird under treatment has a record card, on which symptoms, loss of body mass and treatments are noted. A complete postmortem is performed on each dead bird, and if the carcass is found fresh, organs are fixed, frozen, and bacteriological, virological, histological, and parasitological samples are taken.

A summary of the causes of mortality, based on the records of 258 postmortems between 1987 and 1993, is presented in Fig. 8.1. Proportions of the various etiologies observed in 1989 were the same as in 1987. The increase in the number of deaths in 1989 compared to 1987 was mainly due to the increase in the total number of birds. This was followed by a decrease in mortality rate in 1990 and 1991. The large increase in the total number of birds in 1992 and 1993 explains the high number of deaths. Infectious diseases were the major cause, responsible for 50 per cent to 60 per cent of deaths.

After 1990, the decrease in infectious deaths was due to the eradication of chlamydiosis. In 1993, 'trauma-shock' became the primary cause of death in the captive breeding and rearing units, mainly occurring during visits of 'unknown' people and while birds were being handled. The houbara is very fragile, and must be handled with extreme care by immobilizing the

## Pathology and Veterinary Care

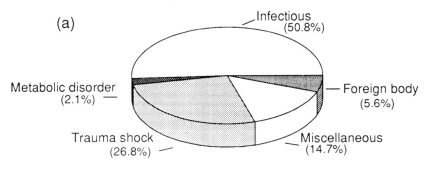

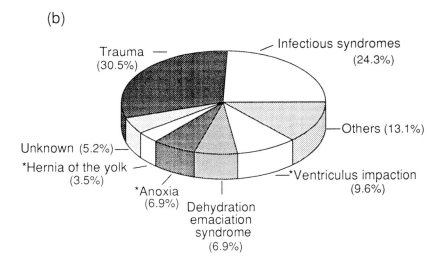

Figure 8.1 Causes of death among (a) houbara between 1987 & 1991 (n = 142), and (b) houbara in 1993 (n = 115)

entire body, with the wings and legs folded against the body (see Chapter 3). Houbara are also highly strung, and deaths or severe injuries can occur when a disturbance frightens the birds in their cages. Cranial traumas inducing death have been observed, as have various fractures and large, open wounds at the base of the neck, due to scraping along the wire mesh. Ingestion of foreign bodies (screws, pieces of wire, fragments of glass, rags) has been recorded on several occasions, inducing severe wounds, and obstructing the gizzard or small intestine, causing death after several days. Particular care must be taken to remove such hazards, as they represent 10 per cent of the total causes of deaths. Other causes of deaths (aneurysm rupture, complete cloacal prolapse) occurred only infrequently.

Mortality of young birds (<1–yr-old) was higher than at other ages. Analyses revealed that mortality occurred mostly during the first month after hatching. Very early mortality (<8 days old) was most common; in 1993, 61 per cent of early deaths occurred before 8 days of age. Causes of these very early deaths were related to infectious and neonatal syndromes. After 1 month of age, most mortality was due to accidental traumas, with only a few birds dying from infectious diseases. Among other pathologies, the most frequent were infectious diseases, ventriculus impaction and secondary hyperparathyroidism.

## 8.2 Infectious Diseases

### 8.2.1 *Chlamydiosis*

Between 1987 and 1989, a so-called 'enteritis-peritonitis' syndrome of unknown etiology killed about 40 birds. This syndrome affected birds sporadically, with a low morbidity, but a high mortality and an acute course (Greth *et al.* 1990). After extensive analysis, it was discovered in 1990 that these mortalities were due to *Chlamydia psittaci*, an intracellular bacterium known to be responsible for chlamydiosis.

*C. psittaci*, the causative agent of psittacosis/ornithosis, occurs worldwide in avian species, generally as a latent infection but also as a clinical infection (Page & Grimes 1984; Eamens & Cross 1989). One reason for the latency seems to be a specific adaptation of this microorganism to the host species (Gylstorff *et al.* 1984), while retaining at the same time its infectiousness with regard to all other possible hosts. Therefore, clinical disease may either develop by passage into another host species or within the same host species, and is generally associated with stress factors (Eamens & Cross 1989). Chlamydial infections were reported by Burkhart and Page (1971) in 15 avian orders, including 140 species. Since then, more species have been shown to be susceptible to psittacosis.

In houbara the disease was of peracute to chronic course.

Morbidity rates among recently arrived birds reached 82 per cent, and the mortality rate was 12 per cent. The following description of the clinical signs should help the practitioner, as well as the breeder, to recognize the disease in houbara bustards.

Clinical signs differ slightly from those that have been described in other birds. Prostration, depression, anorexia and ruffled plumage were the most common and earliest signs. Greenish droppings were observed in the capture pens, and birds had soiled vents due to diarrhoea. Respiratory signs were less often noted. Only upper respiratory symptoms were observed; some animals had a cough, and eleven animals had a translucent mucoid oculo-nasal discharge and signs of tracheitis. Two birds were affected by paralysis of both legs.

A complete postmortem was performed on each dead bird, and if the carcass was found fresh, organs were fixed and frozen. Most of the birds exhibited liver abnormalities: the liver was an unusually dark colour, as if it had been cooked, and of pulpy appearance. The spleen was usually but not systematically enlarged and congested. Colitis, jejunitis and peritonitis were frequently observed. Respiratory lesions were less common: some birds showed tracheitis with abundant mucous obstructing the glottis, and congested lungs. One bird displayed an aerosacculitis.

Further investigations of the histo-pathology and bacteriology focused on the multifactorial etiology of the disease, and revealed some interesting information. The liver displayed signs of a disturbed circulation (hyperaemia, oedema, small haemorrhages and bile congestion), as well as a degree of disseminated necrosis of hepatocytes with detachment from the basement membrane. Within the sinuses an increased number of Kupffer's stellate cells was observed, arranged in the form of a string of pearls. The spleen revealed a depletion of lymphocytes, mainly in the white pulp, and necrosis of some reticular cells (the type could not be established). Lymphatic follicles were rare or non-existent. The walls of many arterioles were either homogenized or invaded by mononuclear cells, in the latter case with activation of the endothelial cells. In a few cases, either hemosiderin was detected, or the number of heterophils in the tissue was markedly reduced. The heart showed a high degree of oedema which also affected the walls of the blood vessels, and a myodegeneratio cordis. In some cases there had been mild infiltration of mononuclear cells into the interstice. In one bird a purulent epicarditis was evident. In the kidney the signs of circulatory disturbance were prominent (hyperaemia, oedema including the walls of the blood vessels, haemorrhages, thrombi). In addition, a necrosis of the tubular epithelium was seen (with or without detachment from the basement membrane or casts in the lumina).

In some birds lesions resembling glomerulonephritis were evident (activated layer of Bowman's capsule, and a non-identifiable eosinophilic staining material within the lumen of the same capsule), in addition to homogenization of the mesangium and widening and thickening of the capillary loops. The modified mesangium consisted mainly of collagenous fibers (van Gieson staining). The lungs showed hyperaemia and oedema, particularly in the atria, and small haemorrhages. There were thrombi mainly in the larger veins. The main lesion was probably an angiosis. The walls of the larger blood vessels displayed a cellular proliferation of mainly mononuclear cells. In the lumina of the larger bronchi there was detritus but no erythrocytes.

The trachea of one bird showed a haemorrhagic tracheitis with loss of the cilia and proliferation of mucous-producing cells. Other tracheae were difficult to evaluate because of mineralized tracheal rings. The mucosa appeared thin compared to other avian species, but cilia were present to some extent. No appreciable numbers of inflammatory cells and lymphatic follicles were demonstrable.

Finally, in the birds described here, and from others with antibodies against *C. psittaci*, it was noted that a peritonitis fibrinosa or fibrosa, particularly affecting the serosa of the intestinal tract, as well as parts of the intestinal wall, was regularly encountered. The disease process apparently began on the serosal side of the intestinal wall. The intestinal mucosa could not be thoroughly evaluated because of autolysis. However, there were no subacute or chronic inflammatory lesions.

From the clinical signs and pathological lesions described here, it is likely that *C. psittaci* contributed to the enteritis-peritonitis syndrome. The clinical, pathological, and histological lesions seen in chlamydiosis are highly variable according to strain and bird species (Gerlach 1986a). Although there are no pathognomonic macroscopic or histological lesions in chlamydiosis (Graham 1989), a comparison of the lesions described here with those of other bird species reveals some unusual features. Although aerosacculitis was observed in only one bird, there was subacute to chronic peritonitis, chronic glomerulitis, and clinically, a flaccid paralysis.

### Table 8.1 Macroscopic Lesions in Houbara Showing Inclusion Bodies Resembling *Chlamydia*

|   | Pneumonia | Peritonitis | Aerosacculitis | Enteritis |
|---|---|---|---|---|
| A | 3/6 | 1/6 | 1/6 | 2/6 |
| B | 50% | 17% | 17% | 33% |

|   | Pericarditis | Splenomegaly | Hepat. Invol | Tracheitis |
|---|---|---|---|---|
| A | 2/6 | 3/6 | 3/6 | 1/6 |
| B | 33% | 50% | 50% | 17% |

A: Number of cases showing the lesion / total number of cases.
B: % of cases showing the lesion / total number of cases

A mild fibrinous aerosacculitis is regarded as an indication for psittacosis/ornithosis in Psittaciformes and Columbiformes (Gerlach 1986a). In houbara, upper respiratory symptoms were prominent (as in Columbiformes), and a haemorrhagic tracheitis was seen in one case. The latter is not assumed to have been caused by *C. psittaci* alone, because there is no report in the available literature. Also, no decision can be made as to whether or not the fibrinous peritonitis was caused by *C. psittaci* and represents a distinctive form of serositis in this avian species. In the so-called enteritis-peritonitis syndrome, there was also no distinct inflammation of the mucosa, but, rather, an infiltration of the wall with inflammatory cells, starting from the serosal side. Another point in favour of the peritonitis not being caused primarily by *C. psittaci*, was that most of the birds died acutely, and the lesions appeared to be subacute to chronic. In any case, we need more information in order to evaluate some of our unusual findings. Either forms are specific to houbara, or they are caused by an incompletely understood multifactorial etiology.

We also observed that houbara were infected by a variety of other viral and bacterial agents: *Serratia marcescens*, *Staphylococcus sp.* (Greth *et al.* 1990). Thus it has to be assumed that the cases described here are caused by a multifactorial etiology, whereby *C. psittaci* is one of the more important agents. Andral *et al.* (1985) described an outbreak of rhinotracheitis in turkeys, with the involvement of *C. psittaci* and of viral agents (adenovirus - hemorrhagic enteritis virus- and paramyxovirus II).

*Enterobacteriaceae* were isolated in high numbers from the dead birds. Since the houbara does not have *Enterobacteriaceae* as a normal component of the gut flora (Greth *et al.* 1990), bacteria of this family are considered potential pathogens, able to cause disease and death following generalization. Death may be due to septicaemia. *C. psittaci* infection was proved to modulate the host immune response (Lammert & Wyrick 1982), and thus to favour potential pathogenic bacteria. The *Enterobacteriaceae* in the cases described here may be the cause of purulent pericarditis, and of the development of thrombi during life. Gram-negative bacterial walls, particularly those of *Enterobacteriaceae*, can activate the alternative pathway of the complement cascade, resulting finally in thrombi (Tizzard 1987). The increase of body mass following treatment for *Enterobacteriaceae* indicates that these bacteria were part of the disease process. The therapeutics used were not particularly effective against *Chlamydia*.

Diagnosis and demonstration of *C. psittaci* in the bustard organs was done by cytology. Impression smears of organs (particularly the spleen) were repeated on the same slide, and on five different slides. A Köster Stamp staining (fuschin and methylen blue) allowed microscopic observations of many intracytoplasmic inclusions morphologically resembling *Chlamydia* (Giroud & Captoni 1964).

The second step of the diagnosis was performed through serological analyses. Blood was taken by alar venipuncture and the extracted sera were kept at $-30°C$. Serological examination was performed with the commercially available competitive enzyme immunoassay Chlamydia-psittaci-AK-EIA (Röhm Pharma GmbH, Darmstadt, Germany). It was used to detect *C. psittaci* antibodies in sera of birds according to the instructions of the manufacturer. Results revealed that the antibody titre was high, and a significant seroconversion was observed in newly imported birds.

Although *C. psittaci* was not identified by culture, the demonstration of intracytoplasmic inclusion bodies morphologically different from the ones caused by *Rickettsia*, the occurrence of antibodies against *C. psittaci*, and the epidemio-clinical findings provided safe grounds for the diagnosis of *C. psittaci*. The antigen used showed cross-reactions only with *C. trachomatis*, an infectious agent which does not occur in avian species.

To maximize the survival chances of houbara being released as part of the reintroduction programme, and to avoid the risk of spreading *C. psittaci* to the indigenous wildlife (Brand 1989), it was decided to attempt to eradicate *C. psittaci* from the breeding flock by means of therapeutic and sanitary measures (see sanitary measures). Tetracyclines are recommended, or even prescribed by law for the treatment of *C. psittaci* infections in several countries. Since psittacosis is a zooanthroponotic disease, and may be transmitted from Psittaciformes to humans, much information has been collected from the Psittaciformes as well as from the more widely kept species. Tetracyclines are considered cross-sensitive with regard to bacterial sensitivity, but in terms of pharmacology, many differences have been observed between the various derivatives. The superior lipid-solubility of doxycycline (DC) provides a good bio-availability and diffusion into the cells, where the initial bodies of *C. psittaci* develop (Gylstorff 1987b). When administered intramuscularly (IM; only possible with the European and Canadian preparation) at 75–100 mg./kg. body mass (BM), efficient blood levels are reached, and sustained for six to seven days (Jakoby 1979a; Jakoby 1979b; Flammer 1989). In contrast to the other tetracyclines, DC is excreted in the faeces without impact on the physiological intestinal flora (Huber 1988). Another side-effect of all antibiotics and, in particular the tetracyclines, is immunosuppression. The possible influence on the immune system should not be ignored, particularly since the microbiological agent is only inhibited by the treatment, but still has to be eliminated actively by the immune system (Gerbermann & Pauels 1982). Before treating all the birds systematically with doxycycline, a pharmacokinetic study was performed to adapt the dosage and to be sure that the minimal therapeutic blood concentration was reached (Greth *et al.* 1993). The mean concentration of doxycycline in the blood was measured after the first injection during the initial seven-day time period, and then again after the seventh injection (each injection administered one day apart), and during seven days following this (Figs. 8.2 & 8.3). Conclusions of the trial are that, due to a rather high plasma peak level, the dosage should be about 80 mg./kg. BM, and SC administration might be generally more desirable. Seven injections (100 mg./kg. BM) are given over a period of 38 days at the following intervals: seven, seven, seven, six, six, five days. Following this treatment, epizootie of Chlamydiosis in the Center has stopped.

Newly arrived birds at the Center have certainly acted as carriers and triggered outbreaks of Chlamydiosis. No mortality attributed to infectious agents occurred from the beginning of 1989 until September 1990. A latent infection had perhaps existed for years in the breeding unit, and may have conferred a premun-

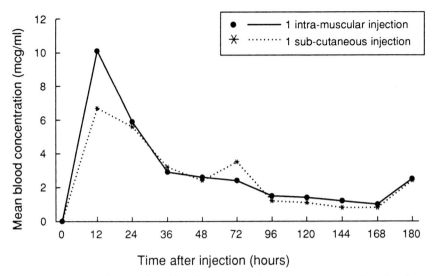

Figure 8.2 Mean doxycycline blood concentrations versus time after the first injection, during the initial seven-day period; intra-muscular (n = 10) and sub-cutaneous (n = 4)

ity status to the birds (Gerlach 1986a). Since a latent infection of *C. psittaci* can be triggered into a manifested clinical disease by various stressing agents, the transport of birds to protected areas (transport itself, change of environment, feed and temperature), should be considered as a significant stress factor.

Houbara are opportunistic feeders that peck the ground frequently, even though food is presented on plates at the NWRC. This behaviour may enhance the transmission of disease in dust and dry particles contaminated by faeces (Brand 1989). Chicks are bred from artificially incubated eggs, in separate rooms without contact with birds of the breeding unit; therefore, it is possible that vertical transmission of *C. psittaci* in the houbara may occur. Such transmission has been shown in ducks, parakeets, gulls and turkeys (Rüffle 1962; Gylstorff 1987a; Eamens & Cross 1989; Olson 1990).

As some bustards are destined to be released in protected areas of Saudi Arabia, an important question is whether or not these birds are suitable for life in the wild. While clinically sick birds can be treated successfully (Meyer & Eddie 1955; Gylstorff et al. 1983; Wachendörfer et al. 1985), an intact immune system is essential (Gerbermann & Pauels 1982), and latent infections cannot be eradicated by medication, as has been shown by Sch-

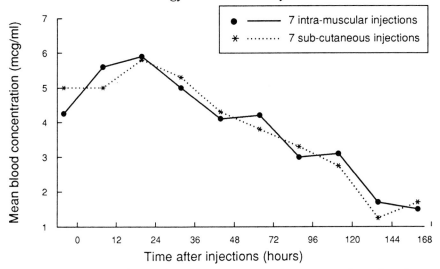

Figure 8.3 Mean doxycycline blood concentrations versus time after the seventh injection, and during the next seven days: intra-muscular (n = 10), sub-cutaneous (n = 4)

achter *et al.* (1978) and Gerbermann *et al.* (1990). Some authors (Shewen 1980; Gerbermann *et al.* 1990) consider a latent infection with avian strains of *C. psittaci* to be normal; they found *C. psittaci* antigens in 31 per cent to 41 per cent of cloacal swabs and faecal samples taken from 119 hawks (Accipitriformes), falcons (Falconiformes), and owls (Strigiformes). Antibodies were demonstrated in 71 per cent of birds kept in a large aviary and used for demonstrations of falconry. Another indication for continued latency is the fact that antibody titres remain high following treatment (Schmeer 1983; Janeczek 1989).

Surveys for chlamydial infections (Burkhart & Page 1971; Ruppaner *et al.* 1984) in wild birds have shown many species to be infected, but with major differences in prevalence between species and regions. Since latency is dependent on the activity of the immune system and environmental stressors (Brand 1989), an outbreak of sub-lethal effects (Brand 1989) could occur in a newly introduced population. However, a serological survey of common avian species in the area, such as ravens (*Corvus ruficollis*) and house sparrows (*Passer domesticus*), has shown that free-living birds have antibodies against *C. psittaci*. This suggests that there is no reason why houbara should not be released into the wild in Saudi Arabia.

## 8.2.2 Other Bacteriological Findings

Analyses of faecal samples show that the aerobic intestinal flora consists mainly of Gram-positive organisms, and that *Enterobacteriaceae* are not normally a component of the faeces. With respect to Gram-negative rods other than *Enterobacteriaceae*, we do not have enough data yet. All isolates of *Enterobacteriaceae* represent facultative pathogens indicating colonization, but also secondary infection or disease and death. Various beta-haemolytic and non-haemolytic *Bacillus* species, as well as various alpha-haemolytic and non-haemolytic *Streptococcus* species, mainly belonging to Lancefield Group D, are isolated from normal intestinal contents. Other specimens found in faecal samples are *Citrobacter amabonaticus* and *Acinetobacter calcoaticus*. Their pathogenicity is not certain.

*Escherichia coli*, *Klebsiella pneumoniae*, *Serratia marcescens*, *Enterobacter*, *Proteus* and *Citrobacter* are considered as secondary invaders, and can be the cause of septicaemia followed by death. This seems to have occurred during a chlamydiosis outbreak. *E. coli* is the most common secondary invader isolated at the Center. Three types have been identified, all fully sensitive to polymixin B, enrofloxacin and furazolidon; $O_1K_1$, $O_2K_1$ and $O_{78}K_{80}$. Chloramphenicol, neomycin and tetracyclins were less efficient. *Klebsiella pneumoniae* strains were usually also fully sensitive to polymixin B, enrofloxacin and furazolidon. Only one salmonella, *Salmonella mbandaka*, an African species, has been found, but in a bird killed by a raptor and without any macroscopic lesions at postmortem. Other bacteria isolated on single occasions are: *Pasteurella multocida*, *Citrobacter freundii*, and *Moraxella*. *Pseudonomonas aeruginosa* is also a frequently isolated secondary invader that can cause death, probably because of toxin production. Type $P_3$ was frequently isolated: this extremely resistant type is dangerous in nosocomial contamination. None of the numerous micrococcaceae isolated demonstrated any clumping, confirming that they are secondary invaders. During the outbreak of Chlamydiosis, many bacteria which were thought to be secondary invaders were isolated. *Serratia marcescens* was frequently isolated and was sensitive to erythromycin and to neomycin. Bacteriological cultures of the major organs revealed *Escherichia coli* $O_{78}K_{80}$ in two birds; *E. coli* $O_1K_1$ in one bird, and *Klebsiella pneumoniae* in two birds.

To evaluate the epidemiological environment and exposure of

houbara to *Mycoplasma gallisepticum* (MG), *Mycoplasma synoviae* (MS), and *Salmonella pullorum* (SP), a serological survey was carried out on the entire flock. No antibodies against MG were detected; serological prevalence to MS and to SP were respectively 2.8 per cent and 3 per cent.

### 8.2.3 Avian Poxvirus

Poxvirus infection in birds is a slowly spreading viral disease, inducing cutaneous lesions on the unfeathered parts of the skin, and/or diphtheric lesions in the digestive and upper respiratory tracts (Tripathy & Cunningham 1984). Natural poxvirus infections have been reported in more than 60 species of wild birds from about 20 families (Clubb 1986).

Avian pox has always occurred in the breeding unit as a mild latent infection in three different clinical forms: cutaneous, diphteric and coryza. Expression in adults was mostly cutaneous, in the form of nodules on the unfeathered parts of the body; legs, toes, or around the oral cavity.

Epidemio-clinical observations showed that three different forms of the disease were present at the Center: in spring 1989, eight chicks died with severe diphteric lesions of the buccal cavity. Consistent clinical signs were anorexia, weakness, and death occurring within a few days (one to 10 days after the discovery of the lesions). Morbidity and mortality rates reached 45 per cent. Chicks seemed to be particularly affected by the disease between the fifth and sixteenth week. The diphteric lesions were white opaque nodules, slightly elevated, developing on mucous membranes. In some of the chicks the nodules coalesced and became large proliferative caseous lesions, with diphteric membranes and pus, invading most of the mouth and covering the tongue. At the end of the progression of the lesion the necrotic membranes could be removed, leaving bleeding erosions. One chick had 21 nodules in the buccal cavity, including the tongue, as well as 14 ulcerations on the oesophagus, and a few circular opaque white nodular lesions on the proximal part of the trachea. Small hemorrhagic spots were also present on the walls of the trachea. Houbara with the cutaneous form had dry scabs around the nares or on the skin. One bird had small ulcers on the legs and the toes, was limping, and had swelling around the left eyelid; prognosis was poor. All the dead birds were frozen in order to carry out virus isolation.

Seven birds of the 1992 generation (between 10 and 14 months old) and three older birds exhibited cutaneous lesions (tibiotarso-tarsometatarsal joint, carpal joint and toes). Lesions were elevated, closed, warm and relatively smooth when palpated. Development of the disease was subacute to chronic, with progressive swelling of nodules. After two weeks, the necrotic tops of the nodules could be removed, leaving bleeding and swelling erosions. After one month, lesions were reduced to dry scabs. Morbidity in the breeding unit remained low during the season, and only the cutaneous form was observed. A significant weight loss ocurred. Prognosis was very good, although one bird displaying a complete fibrosis of the joint had to have this amputated.

Finally, six birds translocated to a reserve for reintroduction in spring 1993, exhibited respiratory symptoms in summer 1993. During the first five days symptoms were only in the upper respiratory tract, i.e. acute sinusitis, conjunctivitis, respiration with a slightly open mouth. Weight loss was progressive. Despite antibiotherapy associated with a symptomatic treatment, the disease reached the lower respiratory tract; dyspnea and open mouth respiration was evident, and rattling and aerosacculitis sounds were noted during auscultation. Endoscopy of one bird revealed an aerosacculitis. Sinusal injections of antibiotics as well as fumigations had variable results. After six days four birds had died.

Electron microscopy confirmed the diagnosis of an avipoxvirus of the family Poxviridae (Joubert 1985) in sampled chicks. For viral culture, samples were inoculated after filtration on the chorio-allantoic membrane of 10–day-old embryonated chicken eggs (Cunningham 1973). Negative staining on both samples, and chorio-allantoic membranes of inoculated eggs, showed typical pox viral particles. A poxvirus was also isolated from two dead embryos, one dying at eight days' incubation and the other of unknown age. This raises the question of the possible transmission of the virus to the eggs. Viral cultures as well as microscopy confirmed the presence of avipoxvirus in the cutaneous lesions of subadult and adult birds sampled at the Center. The strain was isolated and cultivated, and the virus was identified as a member of the Poxviridae family. Further investigations revealed that the virus was not related to known poultry strains. Seroagglutination analyses indicated no antigenic proximity with chicken and turkey pox strains. A noticeable agglutination occurred with a canary pox strain.

Histology also yielded interesting results. In one chick, Bollinger bodies were found in the epithelium of the buccal cavity, towards the larynx. In three other chicks the epithelium of the oesophagus showed a ballooned degeneration with Bollinger bodies. Examination of the spleen in the eight chicks revealed a severe depletion of lymphocytes of the white pulp. Cutaneous biopsies of subadult and adult birds showed a ballooned degeneration and Bollinger bodies.

In the case of the translocated birds, a severe necrotizing pneumonia was found. The tracheal epithelium was very irregular, and sometimes a cellular inclusion body was seen. Small vesicles, most likely the beginning of the ballooning process, had also formed. A focal acute necrosis of tubular cells was observed in one kidney sample. Despite observing no Bollinger bodies, histological signs as well as epidemiology were strongly in favour of a poxvirus origin. The secondary infection seemed to have partially masked the primary causative agent.

In a serological survey done in December 1989 on 33 birds, 18.2 per cent of the birds (six individuals) had antibodies. Of 180 birds tested (in January 1990), only 4.4 per cent (eight individuals) showed antibodies against fowlpox. In the serological survey at the end of January, none of the six birds that had antibodies in December had them in January. In 1990, out of the 45 egg yolks of non-fertile eggs collected, four (8.9 per cent) had antibodies against fowlpox. Avian poxvirus does not produce humoral antibodies to any great extent, which is why the few birds with antibodies against fowlpox are not significant. Serological surveys of the flock are not useful, as seroconversion is not systematic, and the virus strain is not cross-reacting.

No deaths could be directly attributed to avian pox in the adult flock, despite adults dying with severe proliferative lesions in the buccal cavity, and numerous diphtheritic membranes on the oesophagus. Instead they were affected by an enteritis-peritonitis syndrome or a necrotizing pneumonia, due to other infectious agents.

Of the four chicks dying of septicaemia or enteritis, three certainly had a concurrent disease. Intranuclear inclusions due to an unidentified virus were found in the pancreas, the liver or the spleen. Moreover, as shown by histological examination of the spleen, all eight chicks seemed to have severe immunosuppression. It is thus difficult to know the real significance of the pox infection in relation to the other virus. Deaths were the result of secondary bacterial infections, as shown by the bacteriological cultures obtained from the organs. Pneumonia due to *Pseudomonas* was the most usual complication of poxvirus infection. *Escherichia coli* ($O_{18}K_{80}$) as well as *Klebsiella pneumoniae* were described as causal agents of associated enteritis.

There is no specific therapy against avian pox (Gerlach 1986b), although supportive treatment can be carried out. Administration of vitamin A (2000 UI/kg) and antibiotics can be given to support epithelium growth and prevent secondary bacterial infections. Application of iodine and glycerin on the diphteric lesions is ineffective.

In 1993, all sick birds had been previously vaccinated with chicken and turkey vaccinal strains, proving the low efficiency of these strains. Furthermore, despite a good local post-vaccinal reaction (vaccinal take), results showed only a low seroconversion (two birds out of 12 vaccinated became positive). Although immunity against pox is known to be mostly of cellular nature, and poor seroconversion of little prognosis value, the poor result of vaccination against the virus strain of houbara was confirmed by immunotyping (seroagglutination) processed on the isolated strain. According to our present knowledge, adequate immunization would require the use of an autovaccinal strain, or a canary pox strain. Further studies should be carried out in this direction.

Poxvirus infections may occur in stable, captive avian populations if the disease is regularly brought in by free-ranging wild birds or insect vectors (Clubb 1986). Avipoxvirus is not capable of penetrating intact epithelia (Gerlach 1986b); other vectors such as biting insects are necessary. Mosquitos are present during the period of the year when an outbreak occurred in the houbara chicks (June-July), and could be responsible for transmission from bird to bird (Tripathy & Cunningham 1984). It has been shown that *Culex nigripalpus*, once infected by a donor bird, can remain infected for several weeks and thus mechanically transmit the virus (Akey *et al.* 1981). The breeding unit was also visited daily by hundreds of house sparrows, which fed from the houbaras' plates. Many were found dead in the cages and in the alleys with obvious poxvirus lesions in the mouth and around the beak. Since houbara are newly introduced into the area, and since no virological study of the 'houbara strain' has been performed, it is possible that sparrowpox strain is involved. To reduce the risk of sparrows being disease carriers, the entire breeding unit was covered with fine mesh wire to prevent them from entering the cages. The subsequent decrease in the incidence of pox could be explained by the isolation of the houbara from the sparrows, as the vaccination programme of the females and chicks was not fully effective. In 1992 and 1993 only cutaneous forms were observed at the Center. The coryza form of the disease found in young translocated birds seems to be linked to a stress-related immunomodulation (new environment, different weather conditions, different food). Interference of an immunosuppression agent is also suspected.

### 8.2.4 Other Viruses

As in other birds, houbara can be hosts to a number of viruses. Cultures of viruses from frozen samples, as well as electronic microscopy on ultra-thin sections of organs, demonstrated one Herpesvirus, one poxvirus, one Gumboro Disease Virus, three Paramyxovirus (Newcastle Disease Virus, PMV-Pigeon virus and PMV-2) and probably Enterovirus. The strains were sent to specialized laboratories for typing. It is difficult to assess the pathological significance of these findings. Most of the birds showed severe cellular immunosuppression on histological examination of the spleen and blood smears, perhaps due to the Herpesvirus.

To evaluate the epidemiological environment and the exposure of birds to 10 selected viral diseases, as well as to implement an efficient vaccination programme, a serological survey was carried out on a randomly selected group for the following pathogens: paramyxovirus 1 -fowl strain- (NDV), paramyxovirus 1 -pigeon strain- (PMV1P), paramyxovirus 2 (PMV2), paramyxovirus 3 (PMV3), Hemorrhagic Enteritis Adenovirus (HE), celo-adenovirus (CA), Herpes virus -Marek disease- (MDV), Pigeon herpes virus (PHV), Birnavirus -Gumboro disease- (IBD), and Fowlpoxvirus (FP). The serological techniques used were agar gel precipitation tests for HE, CA, GD, MDV, FP and antibodies and hemagglutination inhibition texts for myxoviruses antibodies.

Antibodies to HE, MDV, PHV, FP were detected, but at extremely low prevalences (Table 8.2). Evidence of ND, PMV1P, PMV2, PMV3, CA and GD was found with low but significant prevalence.

In order to understand the incidence of antibodies against Marek's disease virus, pigeon Herpesvirus, Birnavirus, or the Adenovirus strains, it is necessary to isolate the respective virus strains or find the cause of probable cross-reactions. Some birds

## Table 8.2 Prevalence of antibodies for 10 Microbial Pathogens in Houbara

| Pathogen | n | % |
|---|---|---|
| Newcastle disease | 180 | 7.2 |
| Paramyxovirus I Pigeon | 180 | 6.7 |
| Paramyxovirus II | 110 | 6.4 |
| Paramyxovirus III | 128 | 8.6 |
| Hemorrhagic Enteritis Adenovirus | 163 | 0.6 |
| Celo Adenovirus | 180 | 9.4 |
| Marek Disease | 180 | 2.2 |
| Pigeon Herpes Virus | 180 | 1.1 |
| Gumboro Disease | 180 | 7.2 |
| Fowl Poxvirus | 180 | 4.4 |

n = total number of sera tested
% = number of positive sera as a percentage of total tested

showed antibodies against diseases that are specific to poultry, such as Marek's disease and Gumboro disease. Antibodies against the latter have also been found in pheasants (Louzis et al. 1979) and guineafowl (Adewuyi et al. 1989). This suggests that houbara at the NWRC may have been exposed to the virus from poultry farms in the vicinity, but that they did not express any clinical signs.

Prevalence for all pathogens was lower than 10 per cent. In contrast, prevalences for NDV, PMV1P, PMV3 and PHV were significantly higher. The high occurrence of NDV, PMV1P, PMV3 and PHV suggested that the birds had recently been in contact with other birds, probably pigeons, during transportation. It is not surprising that houbara can be infected by avian Paramyxovirus I, because this virus has a wide host spectrum (Gerlach 1986b), and has been reported frequently in countries where the disease is endemic (Alexander 1988). No deaths were directly related to this pathogen, but three birds showed signs of a nervous disorder during this period: i.e., loss of balance, and when flushed they were unable to fly. Two years later two of these birds still exhibited marked nervous disorders, whereas the other one appeared to have recovered completely. The birds may have been infected with a paramyxovirus lentogenic strain, with minimal pathogenicity. This study is the first to report antibodies against NDV, PMV1P, PMV2, PMV3, HE, CA, MDV, PHV, GD, FP, in houbara bustards.

### 8.2.5 Parasitology

Cestodes were found during postmortems, and a thickening of the duodenal walls was frequently noticed. After histology, sections of cestodes were noticed in the duodenal lumen or within the mucosa as a proliferation of the mucosal epithelium of the small intestine. Granulomas, sometimes containing parasitic bodies, were found under the serosa of the intestine and on the mesentery. At least two kinds of cestodes were found in the intestine: the largest was identified as *Raillietina paroniella*, a species that has ants and flies as its intermediate hosts. The smaller species was probably *Idiogenes otidis*, and occurred in huge numbers.

Four nematode species (*Hartertia rotundata*, *Histiocephalus choristidis*, *Subulura brumpti* and *Heterakis gallinarum*) have been found, although they were encountered less frequently than were the cestodes. Subcutaneous larvae of *Ascaridia* or *Spirurida* were also found on one occasion.

According to previous work (Mikaelian 1993), haematozoa of the *Haemoproteus* genus have been found in houbara, as well as a haemoparasite of the *Babesia* genus. In 1993, we observed for the first time an infestation of four young birds with *Trichomonas sp*. Treatment was successfully performed with Carnidazole. The origin of this very localized infection (four birds in the same enclosure), is not known, although sparrows, which are rare but still present in the unit, could be carriers.

Cestodes remain a problem in the breeding unit because infestation can be heavy, particularly in the chicks, and may result in weakness and deficiencies of essential nutrients. Houbara eat insects in their cages, which may serve as intermediate hosts. The incorporation of drugs against these parasites into the houbara food would seem the most practical solution.

## 8.3 Non-infectious Diseases

### 8.3.1 Shock and Traumas

Shock and traumas are the main causes of death in the breeding flock and in captive-reared birds, accounting for 60 per cent of deaths among old (> 1 year) birds in 1993. It is often difficult to prove the traumatic origin of these deaths, and complete epidemiological and pathological analyses are frequently required.

Birds less than 1 month old rarely die from traumas: the youngest bird dead of trauma was 44 days old. After 1 month of age all birds are vulnerable to traumatic accidents. Particularly susceptible are stressed or poorly domesticated birds, and those which have not had their wing feathers cut. Events that can often precede traumas are visits of 'unknown' people, attempts to catch birds, and handling of birds.

Necropsical analyses can help to diagnose a traumatic origin, but can also be very disappointing. Carcasses are always in good condition, and frequently have a significant layer of abdominal fat. Death is always sudden (no symptoms, no decrease in food intake), and necropsy rarely yields obvious indications of the cause of death. Evidence of cerebral congestion or even haemorrhages are not specific to traumas, and can be found following other etiologies. Sometimes concomitant lesions (fractures, luxations) can lead to a traumatic-origin conclusion. It is important to note the position of the dead bird, and to search for evidence of struggling and convulsions; struggling usually occurs when death is not immediate.

The following lesions have been observed: (1) vertebral fracture; (2) vertebral luxation; (3) cerebral haemorrhages; (4) comminutive fracture of the legs and wings; (5) luxation of the femoropelvic joint; and (6) soft tissue injuries. Diagnosis can be very difficult, especially if symptoms of decrease in food intake were observed previously. When associated traumatic lesions are present, diagnosis is easier. The most common traumatic events are cerebral damages: nystagmus, head uncoordination, abnormal gait and leg uncoordination are frequently observed in these cases. Sometimes ataxia or even paralysis is evident. When birds are found dead or paralysed, differential diagnosis must be done to discriminate between nervous manifestations of some acute infectious diseases, and intoxication.

Treatment of paralysed and ataxic birds is hopeless, and the prognosis for birds in shock is poor. Emergency treatment consists of injections (IV and IM) of short-acting corticosteroids every two hours during 36 hours, and after 12 hours the administration t.i.d of a cerebral vasodilator molecule. Antibiotics are recommended to prevent bacterial invasion (Redig 1993). Symptomatic treatment can be employed as well. Supportive treatment consists of the administration of a vitamin B complex, and of strychnine sulphate during five days. The bird should be force-fed, and attempts to make it defaecate are essential, as cloacal sphincter paralysis can occur. In some cases administration of fluids is also necessary (Redig 1993). Among paralysed birds, recovery rates were low, with a mortality rate of 93.4 per cent in 1993. Treatments of concomitant lesions are carried out according to previously described techniques. Broken wings are systematically amputated. When the humerus is concerned, total amputation is necessary, and we recommend a prophylactic amputation of the other wing to avoid imbalance. Leg fractures are difficult to repair successfully. Simple pinning (direct or indirect) is often useless, as it rarely prevents rotation of bones. Use of neutralization plates is not often possible, considering the size of the bone and its pneumatized structure. Fractured bones often appear at the end of the laying season when the birds' ossature is decalcified. We used external skeletal devices with variable success. Double external fixation using external plastic supports instead of metallic ones, which reduces the weight of the device, has given the best results with closed, diaphyseal fractures of the femur.

A few measures can greatly reduce deaths caused by traumas. Preventing stress is strongly recommended: high stress levels as well as the presence of visitors seem to cause most accidents. Between 1992 and 1993, 40 per cent of deaths due to traumas among old (>1–yr-old) birds occurred during such visits. It is essential that visitors are fully informed of the possibility of disturbance to the birds, and are closely supervised while inside the units.

Modification of the fenced environment is often necessary. The use of 'soft' fences is essential; no deaths due to traumas occurred inside the enclosures with walls made of shade-cloth. Therefore, enclosures should have walls made of shade-cloth, with no posts inside the enclosures or close to the walls, and a tensioned net ceiling at 1.5 m height to prevent birds from attempting to take off. Finally, appropriate management techniques are essential; panic is communicative, therefore there should be no more than two birds per breeding enclosure. Feathers of both wings should be cut at least twice a year, and should be checked when the birds are being handled for other reasons. Attempts should be made to limit the physiological decalcification of laying females. Careful handling of birds is necessary; the bird should wear a hood at all times, and the handler should not attempt to restrain the bird from kicking its legs.

### 8.3.2 Hernia of the Yolk and Anoxia

These syndromes are only found in very young chicks (mortality at 0 – 3 days of age). Hernia of the yolk is a neonatal syndrome directly linked to hatching, which is easy to treat when the hernia is small. Treatment consists of ligature of the yolk, followed by excision and disinfection. Antibiotherapy and rehydration by subcutaneous injections of normal saline are always performed. It is necessary to use sterilized equipment throughout the procedure. After the suture is performed, subcutaneous injections of warm and sterile lactated Ringer's solution is necessary. These injections help supply the chick with fluid lost by removal of the yolk sac. They should be done into the inguinal web of the leg or under the lateral skin of the thigh. Approximately 0.4 ml. of rehydrating fluid, which creates a water 'blister' in the fold of the leg, can be administered. The 'blister' will disappear within a few hours. Discontinue the injections when the chick has regained its vigour and is eating normally. Survival of chicks with partial unretraction is much higher than for chicks with total unretraction. This treatment is less effective when the hernia is voluminous, and septicaemia may occur less than 24 hours after hatching.

Anoxia (or long-hatching syndrome) is also linked to the hatching period, and occurred at the beginning of the hatching season. Newly hatched chicks' eyes failed to open, they were unable to stand, and suffered from lethargy and anoxia. Even if force-fed, these chicks inevitably died. The incidence of this syndrome was reduced when oxygen was added to the atmosphere of the incubators.

### 8.3.3 Ventriculus Impaction and Stomacale Perforation

These occurred in chicks and young birds: death resulted from accidental ingestion of foreign bodies (pieces of wire, nails and screws) or by over-ingestion of sand (Plate 8.1). In very young chicks treatment is hopeless, as ventriculostomies are unsuccessful at this age. Foreign bodies usually perforated the ventriculus, and peritonitis followed the perforation. Medical treatment using mineral oil was successfully performed on birds showing a low level of ventriculus impaction. Gizzard impaction is known to cause high mortality during the first three weeks of life in turkey flocks (Riddell 1991). Affected turkeys are usually emaciated owing to an empty intestinal tract, but gizzards are full of a solid mass of interwoven fibrous material. Houbara also have a well-developed gizzard and require grit in their diets for the grinding of seeds and hard foods (Arnall & Keymer 1975). In houbara chicks the gizzard is mainly impacted with fine gravel, which often extends into the first part of the duodenum and the lower intestine. This impaction results from eating litter which the gizzard is unable to handle. If the grit is too fine, impaction rate is increased. Prevention is aimed at discouraging the eating of litter by young chicks. Chicks are gradually introduced to gravel floors, or reared on gravel-free spaces and moved to normal ground only after meals. Further studies concerning the nutritional requirements of houbara chicks, and their qualitative choice of food and grit will certainly improve rearing efficiency.

### 8.3.4 Secondary Hyperparathyroidism

Calcium is needed for bone mineralization by the eighth day of embryonic life: it is derived initially from the yolk and then from the egg shell (O'Connor 1984). Bone metabolic disorders are a common pathological problem in growing chicks, both in wildlife practice (Redig 1993) and in avian pet practice (Fowler 1980; Wallach & Flieg 1967). Calcium deficiency leads to secondary hyperparathyroidism syndrome (increased secretion of PTH as a compensatory mechanism induced by nutritional imbalances, in this case, low calcium content of the diet). Between 1991 and 1993 we observed chicks with calcium deficiency symptoms, which appeared at three weeks of age; i.e. lameness varying from a slight limp to inability to walk and folding fractures of the long bones. Lesions are typical of a fibrous osteodystrophy. We successfully treated these chicks with intensive calcium and vitamin D3 therapy.

## 8.4 Therapeutics

The drugs listed below are not intended to represent a complete pharmacological arsenal for houbara; however, their choice is based on their observed effectiveness and safety (Table 8.3). No particular intolerance to drugs was found. Use of antibiotics must be preceded by a culture and a sensitivity test. However, because it is often necessary to initiate treatment very quickly, broad-spectrum drugs are recommended. A synthesis of the results from all the antibiograms showed that the most efficient antibiotics *in vitro* were quinolones and/or furan derivatives, or aminoglucosides. Baytril (Enrofloxacine) is presently considered to be one of the safest most effective and broad spectrum of all avian antibiotics, and is an excellent 'first choice' before a diagnosis is made (Remple & Riddle 1991).

Short-acting steroids are occasionally used to combat shock and stress. Vitamin and mineral preparations are used to supplement deficiencies and act as tonics. Gaseous anaesthesia with halothane is certainly one of the safest and most efficient methods. Induction time is short, and recovery fast and of very good quality.

Owing to the lack of muscle mass, oral administration of drugs is most desirable for young chicks. With respect to adult birds, medicating food on a 'free choice' basis is not always effective, as consumption is unpredictable. However, for the parenteral administration of drugs, the risks associated with handling and the resulting induced stress on the birds have negative effects that could interfere with the efficacy of the treatment. Incorporating drugs into pellets could be effective, and should be carried out when general treatment or prophylaxis of the entire flock is required, but associated problems may be manufacturing delays, and the homogeneity of the drug's distribution in feed when small quantities are ordered. Intramuscular injections into the breast musculature are always preferred for individual treatment because they are more accurate, and therefore the treatment is more effective. Intravenous administration may be appropriate for delivering fluids or drugs, or to draw diagnostic blood samples. The most easily accessible vein in a houbara bustard is the *venal basilica*, which runs along the inside of the elbow joint. It is easily visible once the feathers have been wet with alcohol. Access to the vein requires two people, one to immobilize the bird on its back. The syringe operator then grasps the clenched wing of the bustard with the left hand and guides the needle into the distended vein with the right hand.

## 8.5 Sanitary and Prophylaxis Plan

Young birds are vaccinated at 2 months of age with a living strain (Hitchner B1) of Newcastle Disease by eye or nares instillation. They are injected one month later with an inactivated strain (Texas strain) and re-instillation with the living strain. Vaccination revealed significant seroconversion and no adverse side-effects. Adult and sub-adult birds receive a booster instillation of this vaccine once a year (Plate 8.2).

Until recently, chicks were vaccinated at 2 months of age against fowlpox with a turkey and chicken vaccinal strain. However, many vaccinated birds develop the disease despite vaccination, and seroconversion is sporadic, with no statistical significance. Therefore, we discontinued systematic immunization, and are trying to develop an autovaccinal strain.

Deworming is performed twice a year on adult birds: in June and December (after and before the laying season). Young birds are dewormed four times a year. Different drugs have been used alternately to prevent parasites from becoming resistant to drugs (see Therapeutics).

It is very important that sanitary procedures are carried out carefully. Regular examinations of the birds are necessary: before and after the laying season houbara are checked, dewormed and blood-sampled. A basic haematological examination is carried out on certain individuals (blood smear, PCV, Red Blood Cell Count). Whenever houbara die, a necropsy is performed within six hours to collect reliable data on the cause of death, and to be able to prevent outbreaks of potentially contagious diseases. A normal dissection procedure is followed that ensures no contamination or destruction of material.

After numerous deaths due to infection in 1989, the breeding unit was isolated from the outside by a finer covering of mesh. A changing room at the entrance of the breeding unit, a hygiene protocol rotating the usage of the cage blocks, as well as regular disinfection with caustic soda were implemented. Subsequently, infection sources were better controlled, and a greater proportion of chicks were successfully reared without any losses as a result of infectious diseases.

A strict quarantine period was established for any bird entering the NWRC. The purpose of quarantine is to prevent the propagation of parasitic, bacteriologic and viral diseases. These diseases may show no clinical signs in healthy birds that survived

**Table 8.3 Drugs and Dosages Used for Houbara**

| Generic Name | Route | Dose | Frequency | Notes |
|---|---|---|---|---|
| Amoxicillin | IM<br>Oral | 100 mg/kg | BID | Effective drug against bacterial diarrhoea and infected injuries |
| Carnidazole | Oral | 20 mg/kg | SID (4 days) | For trichomoniasis |
| Dexamethasone | IM | 2 mg/kg up to 50 mg/kg | BID | Life-threatening shock, trauma (see 8.3.1) |
| Dimetridazole | Oral | 0.5 ml powder in 10 ml water/kg | SID (5 days) | For trichomoniasis |
| Doxycycline | SC<br>Oral | 10 mg/kg | SID | See treatment of Chlamydiosis |
| Enrofloxacine 5% | IM/SC | 10 mg/kg | BID | Broad spectrum antibiotic. Antibiotic of choice for *mycoplasma* |
| Fenbendazole 2.5% | Oral | 25 mg/kg | Once | Prophylatic drug against round worms |
|  | Oral | 25 mg/kg | SID (3 days) | Treatment |
| Flumequin | Oral | 10 mg/kg | BID | Good antibiotic against opportunistic bacteria |
| Ivermectin | Oral | Dilute the bovine preparation 1:4 with propylene glycol and give 0.15 ml/kg | Once | Effective against most gastro-intestinal worms (not tape worms) |
| Kaopectate solution | Oral | 2 ml/kg | TID | As a soothing effect on the gut |
| Mineral oil | Oral | 1–3 ml/kg | Once | Aid in removal of small foreign objects from the ventriculus |
| Niclosamide | Oral | 200 mg/kg | Once | For tape worms |
| Oxytetracycline (LA) | IM | 200 mg/kg | SID |  |
| Praziquantel | Oral<br>IM | 1/4 tab./kg<br>10 mg/kg | Once<br>Once | For tape worms |
| Trimethoprime 40 mg and Sulphonamide 200 mg (Sulfamethoxypyridazine) | IM | 0.2 ml/kg | BID | Prolonged use can cause kidney damage |
| Rifampicin | Oral | 20 mg/kg | BID | Effective against bacterial diarrhoea |
| Tylosin | IM | 20 mg/kg | TID | For *mycoplasma* |
| Vitamin A, D3, E | IM | 0.2 ml/400g | One every 5 days | For A, D3, E hypovitaminosis |

Intramuscular (IM); once daily (SID); once every 12 hours (BID); once every eight hours (TID)

the infection, but can cause disastrous consequences in birds that have not been in contact with the pathogens. Any bird entering the Center is blood-sampled, vaccinated for Newcastle Disease, dewormed and serologically checked. Birds are not released from quarantine until their serological results are available. Birds positive for Chlamydiosis are allowed to join the flock only after Doxycyclin treatment for seven weeks.

Because it is often very difficult to confirm the appearance of discrete infectious agents in the breeding or rearing units, we follow a selected population sample (based on position in the unit, age, sex, sub-species) serologically to detect possible outbreaks of a subclinical disease. Sera are screened once per month against a panel of poultry diseases, including Chlamydiosis.

## 8.6 Conclusions

Mortalities diagnosed as resulting from Chlamydiosis, as well as positive serological results indicate that *C. psittaci* was widely distributed in the Center. Our eradication programme contributed substantially towards increasing the level of hygiene in the flock.

The clinical signs and microscopic pathological lesions observed in the houbara during avian pox outbreaks are similar to those described for fowlpox in other avian species (Tripathy & Cunningham 1984). The virus has been isolated, but the level of mortality observed suggests this strain is particularly lethal (Clubb 1986).

Regular serological analyses revealed that well-known poultry pathogens can still appear in the flock. Efforts must continue to halt the introduction of contagious infectious agents into the Center. Infectious diseases are no longer the primary cause of death, indicating that we have achieved a high level of hygiene. The next challenge must be to prevent deaths from traumas. Avian species are often extremely vulnerable to stress, and frequently die of stress-related traumas. When contemplating an intervention, the practitioner should consider from a medical point of view whether 'the end justifies the means'.

The systematic collection of samples during houbara postmortems or blood sampling has allowed us to gather a wide range of data in various pathological fields. But this survey of disease problems raises more questions than it yields exact diagnoses or solutions. As no data were available previously on this species, many other fields should be investigated. The analysis of resulting data will provide unique information on this poorly known bird, and is necessary for the success of houbara captive-breeding and reintroduction programmes in Saudi Arabia.

# 9
# Production and Prospects

## M. SAINT JALME

The houbara bustard captive breeding programme was initiated in 1986 at the National Wildlife Research Center, with the aim of supporting reintroduction or restocking projects to establish viable houbara populations in the wild in Saudi Arabia. Successful captive propagation of any animal should consider the species' spatial requirements, dietary specializations, environmental needs (temperature, light cycle, humidity), specific housing requirements and medical problems. Most of this knowledge is usually obtained from detailed studies in the natural habitat. Without it, a captive breeding programme would have to proceed on a basis of trial and error (Kleiman 1980). However, due to the low densities of houbara in the wild in Saudi Arabia, and the extreme shyness of this bird, behavioural observations in the field are limited. Fortunately, some biological studies could be carried out in captivity.

Captive-breeding of houbara has proved to be a difficult task. After an initial year of infertile egg production, breeding success was achieved in 1989 with the production of 17 chicks. In 1990 and 1991, 55 and 49 chicks were produced respectively. In 1992, after six years, the captive population became self-sustaining, with the production of 138 chicks, and in 1993, a significant improvement in techniques meant that 285 chicks were produced.

In 1992, a total of 497 eggs were laid (289 by *macqueenii* and 208 by *undulata*), representing an increase of 185 over that laid in 1991 from a comparable number of females. This is the result of intensive egg-pulling. For example, a female *macqueenii* that laid nine eggs in 1991, laid 26 eggs in 1992, establishing a record for the species. Insemination and incubation techniques also improved, increasing levels of fertility from 14 to 69 per cent. This increase can be attributed to more frequent inseminations, weekly monitoring of the physiological status of females, and systematic control of semen concentration and motility.

In 1993 captive breeding results improved again: 670 eggs were laid by 75 females, with a mean fertility level of 85 per cent; an increase of 16 per cent above that achieved in 1992. Research carried out in 1992 showed that inseminations of more than 10 million spermatozoa should be performed on each female, every four or five days. The hatching rate (65.5 per cent excluding early death) increased by 13 per cent, although embryo mortality is still relatively high at the end of incubation (17 per cent). Despite this substantial increase in hatchability, success from artificial methods is still 19 per cent lower than from natural incubation. By application of the best techniques determined by experimentation in 1993, we should be able to improve hatchability to 70 per cent. Production of chicks increased to 285 chicks.

Since 1987, improvements in sanitary conditions have produced a decrease in mortality; in particular, death owing to infectious diseases has decreased sharply. Adult mortality declined from 34 per cent in 1987 to 5 per cent in 1992 and 8 per cent in 1993.

Stability of our breeding stock is maintained by integrating a proportion of the yearly production. However, we must also produce enough houbara for biological studies and release experiments. Mortality of adult breeders determines the allocation of chicks either to breeding or to release programmes. In 1993, 47 captive-bred chicks were released inside the fenced Mahazat as-Sayd reserve, where different release techniques are being explored (Combreau 1994). From figures of current reproductive output from the breeding unit, and the annual mortality of different age classes of captive houbara, we are able to estimate expected production over the next few years. Because only *macqueenii* can be released in Saudi Arabia, the following figures are calculated for that sub-species.

If 50 female chicks are kept back for breeding purposes every year, and values for fertility (85 per cent), hatchability (60 per cent), and annual mortality of adults (10 per cent) remain the same, then the production of *macqueenii* should increase from 152 chicks in 1993, to 1,000 chicks in the year 2002. Production at this level would allow the release of 100 birds in 1994, 150 in 1995, 200 in 1996, 300 in 1997, 400 in 1998, 500 in 1999 and 600 in the year 2000.

*Production and Prospects*

The African sub-species, *undulata*, should not be released in Saudi Arabia. Currently *undulata* are used for experiments on reproductive biology (mating systems and incubation behaviour), thus freeing *macqueenii* females for higher production. They are also used as surrogate parents for *macqueenii* chicks. Ultimately we hope to release *undulata* within their original range, provided cooperation is established with the North African conservation authorities.

# 10

# Conclusions

P. SEDDON

The NCWCD houbara bustard captive-breeding project at the NWRC forms just one part, albeit a large part, of the programme for the restoration of the houbara bustard in Saudi Arabia. The restoration programme has the long-term goal of securing self-sustaining populations of houbara within a network of managed sites throughout Saudi Arabia. The programme has two components: the conservation of wild houbara, resident and migratory, and the reintroduction into suitable habitat of captive-bred houbara (Seddon 1994).

This book deals exclusively with the clearing of the first hurdle of the captive-breeding/reintroduction phase; houbara can now be bred in captivity. Although a major achievement, and one on which the success of this phase of the programme was dependent, perhaps an even greater challenge lies ahead – the reintroduction of captive-bred houbara back into the wild. First results of 'soft release' trials in the 2,200km$^2$ fenced Mahazat as-Sayd protected area have been instructive, and even encouraging. A small group of houbara have successfully made the transition from captivity to fully independent life in the reserve. However, other released houbara have fallen prey to foxes and raptors; clearly much work is needed to assist predator-naïve houbara to develop effective anti-predator behaviours. Modification of chick-rearing techniques to limit contact with humans and maintain the houbara's innate wariness of non-houbara may be part of the answer. For a relatively long-lived species like the houbara, survival for the first year is probably an important criterion for the success of releases. The evidence so far shows that survival in the reserve for at least the first month may be a good indicator of longer-term survival. The question remains, however, whether released houbara will remain and breed successfully within the

## Conclusions

reintroduction site where they are protected from hunting. More releases are planned, and new reintroduction sites are being developed (Seddon *et al.* 1995).

The conservation of resident wild houbara populations appears at this stage to carry only limited potential for a major increase in houbara numbers in Saudi Arabia. Currently breeding resident houbara are confined to a single locality in the far north of the Kingdom – the ~12,000km$^2$ Harrat al-Harrah protected area. Harrat al-Harrah is not fenced, but is protected by rangers on the ground, and by aerial patrols. However, in the seven years since the reserve was created and hunters, sheep and goats were removed, there has not been any demonstrable increase in numbers of breeding houbara. Breeding and breeding success in the reserve may be constrained by a limited and unreliable annual rainfall.

Saudi Arabia holds not only small resident populations of houbara, but is also the winter home to much larger numbers of migratory houbara which move south into the Kingdom each year after breeding. The location of the breeding grounds of these migrants is not precisely known, but is likely to lie somewhere between Saudi Arabia and the former Soviet Republics of Uzbekistan and Kazakhstan. In good years, that is, years with good winter rainfall and plant growth, it is possible that some of these migrant houbara may remain to breed in Saudi Arabia. It is therefore these migrant houbara which may hold the key to a natural increase in the number of breeding houbara in Saudi Arabia. Conservation efforts must therefore consider not only houbara within the Kingdom, but also seek to protect and encourage the houbara in breeding areas outside Saudi Arabia. To this end conservation authorities in Saudi Arabia are actively promoting an international management strategy for the houbara, and are supporting collaborative projects in countries which may hold houbara populations of importance to Saudi Arabia.

Whether releasing captive-bred houbara, or managing wild populations, there is a need to know more about the basic ecology of the bird, including its status, distribution and movements throughout the entire Arabian Peninsula, the relationship between resident houbara and migrants, and those factors which enhance the reproductive success and survival of houbara in the wild. The success of the captive-breeding project presented here gives us a safety net; we now know we can maintain populations in captivity. But the goal is not to create a self-perpetuating zoo; captive-breeding is only a tool for conservation, one facet in the restoration of the houbara bustard in Saudi Arabia.

# 11
# References

Abouammoh, A.M. 1991. 'The Distribution of Monthly Rainfall Intensity at Some Sites in Saudi Arabia'. *Environmental Monitoring and Assessment* 17: 89–100.

Adewuyi, O.A., O.A. Durojaiye and D.F. Adene. 1989. 'The Status of Guinea Fowls (*Numida meleagridis*) in the Epidemiology of Infectious Bursal disease (IBD) of Poultry in Nigeria'. *J. Vet. Med. B.* 36: 43–48.

Akey, B.L., J.K. Nayar and D.J. Forrester. 1981. Avian Pox in Florida Wild Turkeys: *Culex nigripalpus* and *Wyeomyia vanduzeei* as Experimental Vectors'. *J. Wildl. Dis.* 17: 597–599.

Alekseev, A.F. 1985. 'The Houbara Bustard in the North-west Kyzylkum (USSR)'. *Bustard Studies* 3: 87–91.

Alexander, D.J. 1988. *Newcastle Disease*. Kluwer Academic Publishers, Dordrecht, The Netherlands.

Ali, S. and S.D. Ripley. 1980. *Handbook of the Birds of India and Pakistan*. Vol. II. Oxford University Press, Oxford. 2nd edition. pp. 185–198.

Andral, B., M. Metz, D. Toquin, J. Le Coz and J. Newman. 1985. 'Respiratory Disease (Rhinotracheitis) of Turkeys in Britanny, France. III. Interaction of Multiple Infecting Agents'. *Avian Dis.* 29: 233–243.

Arnall, L. and I.F. Keymer. 1975. *Birds Diseases*. TFH Publications, Neptune, NJ., USA.

Birkhead, T.R. and A.P. Moller (eds) 1992. *Sperm Competition in Birds, Evolutionary Causes and Consequences*. Academic Press, London, New York and Tokyo.

Brand, C.J. 1989. 'Chlamydial Infections in Free-living Birds'. *J. Am. Vet. Med. Assoc.* 195: 1531–1535.

Brillard, J.P. 1982. 'Aspects pratiques de l'insemination artificielle des femelles (poules, pintades, dindes)' in: *Fertilité et insémination artificielle en aviculture*. ed. INRA. Versailles. pp. 77–102.

Burkhart, R.L. and L.A. Page. 1971. 'Chlamydiosis (Ornithosis-psittacosis'. in: Infectious and Parasitic Diseases of Wild Birds. ed. J.W. Davis, R.C. Anderson, L. Karstad *et al.* Iowa State University Press, Ames, Iowa. pp. 118–140.

Clarke, J.E. 1982. 'The Houbara Bustard in Jordan'. *Sandgrouse* 4: 111–113.

Clubb, S.L. 1986. 'Avian Pox in Cage and Aviary Birds'. in: Zoo and Wild Animal Medicine. ed. M.E. Fowler. WB Saunders Company. pp. 213–220.

Coles, C.L. and N.J. Collar (eds) 1980. Symposium papers. 'The Great Bustard. Sofia 1978'; 'The Houbara Bustard. Athens 1979' Sydenham Printers, Poole.

Collar, N.J. 1980. 'Bustards in Decline'. *British Birds* 73: 198–199.

—— and P.D. Goriup (eds) 1983. 'The ICBP Fuerteventura Houbara Expedition: Introduction'. *Bustard Studies* 1: 1–92.

Collins, D. 1980. 'Aspects of the Behaviour and Ecology of the Houbara Bustard (*Chlamydotis undulata fuertaventurae*)'. Unpublished, preliminary report, first field season 1979–1980.

—— 1983. 'Habitats and Vegetation on Fuerteventura'. in: 'The ICBP Fuerteventura Houbara Expedition, 1979'. eds N.J. Collar and P.D. Goriup. *Bustard Studies* 1: 37–42.

—— 1984. 'A Study of the Canarian Houbara Bustard (*Chlamydotis undulata fuertaventurae*) with Special Reference to its Behaviour and Ecology'. M. Phil. thesis, University of London.

Combreau, O. 1994. 'Houbara Releases in Mahazat as-Sayd Reserve (Saudi Arabia): Techniques and First Results'. Unpublished report. NWRC, Taif, Saudi Arabia.

Cramp, S. and K.E.L. Simmons (eds) 1980. *Handbook of the Birds of Europe, the Middle East and North Africa*. Vol. II. Oxford University Press, Oxford. pp. 636–668.

Cunningham, C.H. 1973. *A Laboratory Guide in Virology*. 7th edition. Burgess Publishing Co.

Dawson, A. and G. Bartholomew. 1968. 'Temperature Regulation and Water Economy of Desert Birds'. in: *Desert Biology*. Academic Press, New York. 357–389.

Dement'ev, G.P. and N.A. Gladkov (eds) 1968. 'Birds of the Soviet Union'. Vol. II. 163–195. 1968 translation from Russian by A. Birron, Z.S. Cole and E.D. Gordon. Israel Program for Scientific Translation, Jerusalem.

Eamens, G.J. and G.M.J. Cross. 1989. 'Chlamydiosis' in: *Diseases of Cage Birds*. ed. E.W. Burr. TFH Publications, Neptune, NJ, USA. pp. 114–118.

Emlen, T. and L.W. Oring. 1977. 'Ecology, Sexual Selection and the Evolution of Mating Systems'. *Science* 197: 215–223.

Etchécopar, R.D. and F. Hüe. 1978. *Les Oiseaux de Chine, non passereaux*. Éditions du Pacifique, Papeete (Tahiti). pp. 275–279.

Farner, D.S. and E. Gwinner. 1980 (eds) 'Photoperiodicity, Circannual, and Reproductive Cycles' in: *Avian Endocrinology*. ed. A. Epple and M.H. Stetson. Academic Press, New York. pp. 331–366.

Fowler, M.E. 1980. 'Ossification of Long Bones in Raptors'. in: *Recent Advances in the Study of Raptor Diseases; Proceedings of the International Symposium on Diseases of Birds of Prey, July 1980*. eds. J.E. Cooper and A.G. Greenwood. London. pp. 75–82.

Flammer, K. 1989. 'Treatment of Chlamydiosis in Exotic Birds in the United States'. *J. Am. Vet. Med. Assoc.* 195: 1537–1540.

## References

Gallagher, M. and M.W. Woodcock. 1980. *The Birds of Oman*. Quartet Books, London.

Gaucher, P. 1987. 'Biological Study on the Reproduction of the Houbara Bustard in Algeria, Part II'. Unpublished report. NWRC, Taif, Saudi Arabia.

—— 1988. 'Experiments with Houbara Bustard'. Unpublished report. NWRC, Taif, Saudi Arabia.

—— 1991. 'On the Feeding Ecology of the Houbara *Chlamydotis undulata undulata*'. *Alauda* 59: 120–121.

—— P. Chappuis, P. Paillat, M. Saint Jalme, F. Lotfikhah and M. Wink. In Press. Updated reference 'The Taxonomy of the Houbara Bustard *Chlamydotis undulata*: Subspecies Revised on the Basis of Sexual Display and Genetic Divergence. Ibis.

Gerbermann, H. and F.J. Pauels. 1982. 'Der Einfluss des Immunsystems auf die Abwehr einer Psittakoseinfektion'. *Der Praktische Tierarzt.* 63: 458–460.

—— J.R. Jakoby and J. Kösters. 1990. 'Chlamydienbefunde aus einer grösseren Greifvogelhaltung'. *J. Vet. Med. B.* 37: 739–748.

Gerlach, H. 1986a. 'Chlamydia'. in: *Clinical Avian Medicine and Surgery.* ed. G.J. Harrison and L.R. Harrison. WB Saunders Company, USA. pp. 457–463.

—— 1986b. 'Viral Diseases' in: *Clinical Avian Medicine Surgery.* ed. G.J. Harrison and L.R. Harrison. WB Saunders Company, USA. pp. 408–433.

Giroud, P. and G. Captoni. 1964. 'Diagnostic des Rickettsioses Humaines' in: *Techniques de laboratoire en virologie humaine.* ed. P. Lepine. Masson, Paris, France. pp. 793–811.

Goodman, S.M. and P.L. Meininger (eds) 1989. *The Birds of Egypt.* Oxford University Press, Oxford. pp. 227–230, 484.

Goriup, P.D. (ed.) 1983. *The Houbara Bustard in Morocco – Report of the Al Areen/ICBP, March 1982.* Oriental Press, Bahrain.

—— and J. Norton. 1990. 'Houbara Bustard Field Research Project. Progress report. December 1990'. Unpublished report. NCWCD, Riyadh, Saudi Arabia.

—— and M. Al Salamah. 1992. 'Houbara Bustard Field Research Project: Report for 1990'. Unpublished report. NCWCD, Riyadh, Saudi Arabia.

Graham, D.L. 1989. 'Histopathologic Lesions Associated with Chlamydiosis in Psittacine Birds'. *J. Am. Vet. Med. Assoc.* 195: 1571–1573.

Green, A.A. 1984. 'The Avifauna of the Al Jawf Region, Northwest Saudi Arabia'. *Sandgrouse* 6: 48–58.

Greth, A., H. Gerlach, B. Andral and M. Vassart. 1990. 'Pathology of the Houbara Bustard in Captive Breeding Scheme in Saudi Arabia'. VIth Internat. Conf. Wildlife Dis. Berlin. Abstracts. p. 4.

—— H. Gerlach, M. Gerbermann, M. Vassart and P. Richez. 1993. 'Pharmacokinetics of Doxycycline after Parenteral Administration in the Houbara Bustard (*Chlamydotis undulata*)'. *Avian Disease* 37: 31–36.

Gylstorff, I. 1987a. '*Chlamydia psittaci*' in: *Gylstorff/Grimm Vogelkrankheiten.* Verlag Eugen Ulmer, Stuttgart. p. 319.

―――― 1987b. 'The Treatment of Chlamydiosis in Psittacine Birds'. *Isr. J. Vet. Med.* 43: 11–19.

―――― J.R. Jakoby and H. Gerbermann. 1983. 'Vergleichende Untersuchungen zur Psittakosebekämpfung auf medikamenteller Basis. I. Mitteilung: Bei Amazonen erreichbare Blutspiegel nach oraler Verabreichung von unterschiedlichen Fütterungsarzneimitteln and nach parenteraler Verabreichung von Doxycyclin in einer Importstation'. *Berliner Münchner Tierärztliche Wochenschrift* 96: 262–264.

―――― J.R. Jakoby and H. Gerbermann. 1984. 'Vergleichende Untersuchungen zur Psittakosebekämpfung auf medikamenteller Basis. II. Mitteilung: Wirksamkeitsprüfung verschiedener Arzneimittel bei unterschiedlichen Applikationsformen bei experimentell infizierten Grünwangen Amazonen (*Amazona viridigenalis*)'. *Berliner Münchner Tierärztliche Wochenschrift* 97: 91–99.

Haddane, B. 1985. 'The Birds of Morocco: A Brief Review'. *Bustard Studies* 3: 109–112.

Hall, M.R., E. Gwinner and M. Bloesch. 1987. 'Annual Cycles in Moult, Body Mass, Luteinizing Hormone, Prolactin and Gonadal Steroids during the Development of Sexual Maturity in the White Stork'. *J. Zool. Lond.* 211: 467–486.

Heim de Balzac, H. and N. Mayaud. 1962. *Les Oiseaux du nord-ouest de l'Afrique*. Paul Lechevalier, Paris. pp. 109–113.

Huber, W.G. 1988. 'Tetracyclines' in: *Veterinary Pharmacology and Therapeutics*, eds N.H. Booth and L.E. Mcdonald. Ames, State University Press. 813–321.

Jakoby, J.R. 1979a. 'Versuche zur Prophylaxe und Therapie der Psittakose bei Papageien und Sittichen'. *Berliner, Münchner Tierärztliche Wochenschrift*. 92: 91–95.

―――― 1979b. 'Doxycylin zur Behandlung und Prophylaxe der Psittakose'. Bericht 1. Tagung "Krankheiten der Vögel", München. Hrsg. Dtsch. Vet. Med. Ges, Giessen.

Janeczek, F. 1989. '*Chlamydia psittaci* – Diagnostik bei Psittaciformes: Vergleichende Untersuchungen zum Antigennachweis in der Zellkultur und im ELISA sowie zum Antikörpernachweis in der Komplementbindungsreaktion und im Blocking ELISA'. Veterinärmedizin Dissertation, Müchen.

Joubert, L. 1985. 'Les Principaux poxvirus pathogènes pour les animaux'. *Virologie Médicale*. ed. J. Maurin. Flammarion Médecine-Sciences, Paris.

Kleiman, D.G. 1980. 'The Sociobiology of Captive Populations' in: *Conservation Biology: An Evolutionary Ecological Perspective*, eds M.E. Soulé and B.A. Wilcox. Sunderland, Massachusetts. pp. 243–261.

Lake, P.E. 1975. 'Gamete Production and the Fertile Period with Particular Reference to Domesticated Species'. *Symposium 2001 Soc. Land.* 35: 225–244.

Lammert, J.A. and P.B. Wyrick. 1982. 'Modulation of the Host Immune Response as a Result of *Chlamydia psittaci* Infection'. *Infection and Immunity* 35: 537–545.

## References

Launay, F. and P. Paillat. 1990. 'A Behavioral Repertoire of the Adult Houbara Bustard (*Chlamydotis undulata macqueenii*)'. *Rev. Ecol.* (Terre Vie) 45: 65–88.

Lavee, D. 1985. 'The Influence of Grazing and Intensive Cultivation on the Population Size of the Houbara Bustard in the Northern Negev in Israel'. *Bustard Studies* 3: 103–197.

—— 1988. 'Why is the Houbara *Chlamydotis undulata macqueenii* Still an Endangered Species in Israel?' *Biol. Conserv.* 45: 47–54.

Lodge, J.R., N.S. Fechheimer and R.G. Jaap. 1971. 'The Relationship of "*in Vivo*" Sperm Storage Interval to Fertility and Embryonic Survival in Chicken'. *Biology of Reproduction* 5: 252–257.

Louzis, C., S.I. Gillet, K. Irgens, A. Jeannin and J.P. Picault. 1979. *Bulletin Mensuel de la Société Vetérinaire Pratique de France* 63: 785.

Malik, M.M. 1985. 'The Distribution and Conservation of Houbara Bustards in North-west Frontier Province'. *Bustard Studies* 3: 81–85.

Meinertzhagen, R. 1954. *Birds of Arabia*. Henry Sotheran, London. pp. 544–547.

Mendelssohn, H., U. Marder and M. Stavy. 1979. 'Captive Breeding of the Houbara (*Chlamydotis undulata macqueenii*) and a Description of its Display'. *XIII ICBP Bulletin*: 134–149.

—— 1980. 'Development of the Houbara (*Chlamydotis undulata*) Populations in Israel and Captive Breeding' in: *Symposium Papers*, eds C.L. Coles and N.J. Collar. Sydenham Printers, Poole.

—— 1983. 'Observations on the Houbara (*Chlamydotis undulata*) in Israel' in: *Bustards in Decline*, eds P.D. Goriup and H. Vardhan. Tourism and Wildlife Society of India, Jaipur. pp. 91–95.

—— U. Marder and M. Stavy. 1983. 'Captive Breeding of the Houbara (*Chlamydotis undulata macqueenii*) and Development of Young Bird' in: *Bustards in Decline*, eds P.D. Goriup and H. Vardhan. Tourism and Wildlife Society of India, Jaipur. pp. 288–292.

Meyer, K.F. and B. Eddie. 1955. 'Chemotherapy of Natural Psittacosis and ornithosis – Field Trial of Tetracycline, Chlortetracycline, and Oxytetracycline'. *Antibiotics Chemotherapy* 5: 289–299.

Mian, A. 1984. 'A Contribution to the Biology of Houbara: 1982–1983 Wintering Population in Baluchistan'. *J. Bombay Nat. Hist. Soc.* 81: 537–545.

—— 1989. 'A Contribution to the Biology of the Houbara Bustard: 1983–84 Population Levels in Western Baluchistan'. *J. Bombay Nat. Hist. Soc.* 86: 161–165.

—— and M.I. Surahio. 1983. 'Biology of the Houbara (*Chlamydotis undulata macqueenii*) with Reference to Western Baluchistan'. *J. Bombay Nat. Hist. Soc.* 80: 111–118.

—— and A.A. Dasti. 1985. 'The Houbara Bustard in Baluchistan, 1982–83: A Preliminary Review'. *Bustard Studies* 3: 45–49.

Mikaelian, I. 1993. 'Variations circannuelles des paramètres hématologiques de l'outarde houbara (*Chlamydotis undulata*)'. *Th. Med. Vet. Lyon.*

Mirza, Z.B. 1985. 'A Note on Houbara Bustards in Cholistan, Punjab'. *Bustard Studies* 3: 43–44.

Nalbandov, A.V. and L.E. Card. 1943. 'Effect of Stale Sperm on Fertility and Hatchability of Chicken Eggs'. *Poultry Science* 22: 218–226.

O'Connor, R.J. 1984. *The Growth and Development of Birds*. Wiley-Interscience Publication, Chichester.

Olson, G.H. 1990. 'A Review of Some Causes of Death of Avian Embryos'. *Proc. Ann. Conf. Assoc. Avian Vet.* 106–111.

Osborne, P.E. 1992. 'Preliminary Field Survey of the Houbara Bustard in Abu Dhabi, United Arab Emirates. Phase 2 – Survey of Potential Breeding Habitat in Abu Dhabi, February-March 1992'. Interim report to the Scientific Advisory Panel of the National Avian Research Centre, Abu Dhabi, National Avian Research Centre, Abu Dhabi.

Page, L.A. and J.E. Grimes. 1984. Avian chlamydiosis (Ornithosis) in *Diseases of Poultry*, 8th edition, eds M.S. Hofstad, H.J. Barnes, B.W. Calnek, W.M. Reid, and H.W. Yoder. Iowa State University Press, Ames, Iowa. pp. 283–308.

Paillat, P. 1987. 'Report on the Baluchistan Expedition, Pakistan 1987'. Unpublished report. NWRC, Taif, Saudi Arabia.

Platt, J.B. 1985. 'Houbara Bustard Research in Dubai, United Arab Emirates'. *Bustard Studies* 3: 101–102.

Ponomareva, T.S. 1983. 'Comportement reproducteur et distribution de l'Outarde houbara (*Chlamydotis undulata macqueenii*) sur les lieux de nidification'. *Zool. Zhum.* 62: 592–602 (translated into French; original paper in Russian with English summary).

—— 1985. 'The Houbara Bustard: Present Status and Conservation Prospects (in the USSR)' *Bustard Studies* 3: 93–96.

Porter, R.F. and P.D. Goriup. 1985. 'Recommendations for the Conservation of the Arabia Bustard and Houbara Bustard in Saudi Arabia'. Unpublished IUCN report, Riyadh, Saudi Arabia.

Povlawski, A.N. 1965. 'Zur Biologie der Kragentrappe'. *Falke* 12: 242–243.

Ramadan, G. 1989. 'Breeding the Houbara Bustard at the Al Ain Zoo and Aquarium, Abu Dhabi'. *UAE Zool. Garten* 59: 229–240.

Razdan, T. and J. Mansoori. 1989. 'A Review of the Bustard Situation in the Islamic Republic of Iran'. *Bustard Studies* 4: 135–145.

Redig, P.T. 1993. 'Fluid Therapy and Acid-base Balance in the Critically Ill Avian Patient' in: *Medical Management of Birds of Prey*. ed. P.T. Redig. Saint-Paul, University of Minnesota. pp. 49–60.

Remple, J.D. and S.K. Riddle. 1991. 'Pharmacological Considerations' in *Falcon Propagation*, eds J.D. Weaver and T.J. Cade. The Peregrine Fund Inc. Boise, Idaho, USA. pp. 77–80.

de Revier, M. 1982. 'Differents aspects de l'elevage des reproducteurs males (coqs, pintades) utilisés en insemination artificielle' in: *Fertilité et insémination artificielle en aviculture*. ed. INRA. Versailles. pp. 25–60.

Riddell, C. 1991. 'Developmental, Metabolic and Miscellaneous Disorders'. in: *Diseases of Poultry*. ed. B.W. Calnek, H.J. Barnes, C.W. Beard, W.M. Reid and H.W. Joder Jr. Wolfe Publishing Ltd. pp. 827–862.

## References

Rüffle, E. 1962. 'Über Funde von Ornithosevirus in Hausenteneiern'. *Monatshefte Veterinärmedizin* 17: 879–881.

Ruppanner, R., D.E. Behmeyer, W.J. Delong III, C.E. Franti and T. Schulz. 1984. 'Enzyme Immunoassay of Chlamydia in Birds'. *Avian Dis.* 28: 608–615.

Saint Jalme, M., P. Gaucher and P. Paillat. 1994. 'Artificial Insemination in Houbara Bustards (*Chlamydotis undulata*): Influence of the Number of Spermatozoa and Insemination Frequency on Fertility and Ability to Hatch'. *Journal of Reproduction and Fertility* 100: 93–103.

Saleh, M.A. 1989. 'The Status of the Houbara Bustard in Egypt'. *Bustard Studies* 4: 151–156.

Sauveur, B. 1982. 'Notion de physiologie de la reproduction femelle en relation avec l'insémination artificielle' in: *Fertilité et insémination artificielle en aviculture*. ed. INRA. Versailles. pp. 61–75.

—— 1988. *Reproduction des volailles et production d'œufs*', ed. INRA. Paris.

Schachter, J., N. Sugg and M. Sung. 1978. 'Psittacosis: The Reservoir Persists'. *J. Infect. Dis.* 137: 44–49.

Schmeer, N. 1983. 'Enzymimmuntest zum Nachweis von IgG- und IgM-Antikörpern gegen *Chlamydia psittaci* bei der Taube'. *Zentralblatt Veterinärmedizin B* 30: 356–370.

Schulz, H., M. Schulz, P. Paillat, P. Gaucher and X. Eichaker. 1991. 'Incubation Parameters of the Houbara (*Chlamydotis undulata*)'. Unpublished report. NRWC, Taif, Saudi Arabia.

Seddon, P.J. 1994. 'Species Conservation Strategy for the Houbara Bustard (*Chlamydotis undulata macqueenii*) in Saudi Arabia; revised draft'. Unpublished report, NCWCD, Riyadh, Saudi Arabia.

—— and Y.M. van Heezik. 1994. 'Houbara Bustard Field Research: Annual Report'. Unpublished report, NRWC, Taif, Saudi Arabia.

—— and Y.M. van Heezik. 1995. 'Interseasonal changes in houbara bustard density' in Harrat al-Harrah, 'Saudi Arabia: implications for managing a remnant population'. *Biol. Conserv.*

—— M. Saint Jalme, Y. van Heezik, P. Paillat, P. Gaucher and O. Combreau. 1995. Restoration of Houbara Bustard Populations in Saudi Arabia. Developments and Future Directions. *Oryx* 29: 136–142.

Sharp, P.J. 1980. 'Female Reproduction'. in: *Avian Endocrinology*. ed. A. Epple and M.H. Stetson. Academic Press, New York and London, 435–454.

Shewen, P.E. 1980. 'Chlamydial Infection in Animals: A Review'. *Can. Vet. J.* 21: 2–11.

de Smet, K. 1989. 'The Houbara Bustard in Algeria: A Preliminary Report'. *Bustard Studies* 4: 157–159.

Surahio, M.I. 1985. 'Ecology and Distribution of Houbara Bustards in Sind'. *Bustard Studies* 3: 55–58.

Symens, P. 1988. 'Houbara Bustard Survey in Harrat al Harrah, March 1988'. Unpublished report, NWRC, Taif, Saudi Arabia.

Tizzard, I. 1987. *Veterinary Immunology.* 3rd edition. W.B. Saunders Company, Philadelphia, London, Toronto, Sydney, Tokyo and Hong Kong.

Tripathy, D.N. and C.H. Cunningham. 1984. 'Avian Pox' in *Diseases of Poultry.* 8th edition, eds M.S. Hofstad, H.J. Barnes, B.W. Calnek, W.M. Reid and H.W. Yoder. Iowa State University Press. Ames, Iowa. pp. 524–534.

Urban, E.K., C.H. Fry and S. Keith (eds) 1986. *The Birds of Africa.* Vol. II. Academic Press, London. 148–179.

Vanderheyden, N. 1989. 'Hematology and Plasma Chemistry Values in Selected Diseases of Amazon Parrots'. in: *Sec. Europ. Symp. Avian Med. Surg., March 8–11.* pp. 357–364.

Wachendörfer, G., W. Lüthgen, J. Brettschneider, B. Burski and J. Kubicek. 1985. 'Erfahrungen mit einer Quarantäne und Chemoprophylaxe bei importierten Papageien und Sittichen im Rahmen der staatlichen Psittakosebekämpfung'. *Deutsche Tierärztliche Wochenschrift* 92: 301–344.

Wallach, J.D. and G.M. Flieg. 1967. 'Nutritional Secondary Hyperparathyroidism in Captive Psittacine Birds'. *J. Am. Vet. Med. Assoc.* 155: 1046–1051.

Weaver, J.D. and T.J. Cade (eds) 1991. *Falcon Propagation. A Manual on Captive Breeding.* The Peregrine Fund Inc. Boise, Idaho.